Ideas into Words

PUBLISHING FOR THE WORLD
125 Years

THE JOHNS HOPKINS UNIVERSITY PRESS

Mastering the Craft of Science Writing

Ideas into Words

Elise Hancock

Foreword by
Robert Kanigel

THE JOHNS HOPKINS UNIVERSITY PRESS   Baltimore & London

©2003 The Johns Hopkins University Press
Foreword © 2003 Robert Kanigel
All rights reserved. Published 2003
Printed in the United States of America on acid-free paper
9 8 7 6 5 4 3 2 1

The Johns Hopkins University Press
2715 North Charles Street
Baltimore, Maryland 21218-4363
www.press.jhu.edu

Library of Congress Cataloging-in-Publication Data
Hancock, Elise.
    Ideas into words: mastering the craft of science writing / Elise Hancock.
       p. cm.
    ISBN 0-8018-7329-0 — ISBN 0-8018-7330-4
    1. Technical writing.   I. Title.
T11 .H255   2003
808'.0665—dc21   2002011065

A catalog record for this book is available from the British Library.

For my father,
who would have been so proud.

Contents

Foreword

As I stepped into her office, I found Elise in her desk chair, bent over a page of manuscript rolled up into her typewriter. She didn't look up. She never looked up. Just a year or two earlier, that would still have infuriated me. *Social graces, Elise? Remember those?* But by now I was long past the point where I paid it any mind. So I sat and waited while she finished.

Finally, she pulled out the page, gathered it together with one or two others and, still not looking up, passed them to me. It was a short essay for the *Johns Hopkins Magazine*, which she edited, but this was one of the little pieces she wrote herself. What, she wanted to know, did I think of it?

Oh, it was fine, I too quickly said after reading it, then paused. I was a freelance writer, of the perpetually struggling sort, had done some assignments for Elise, and sought others. Elise was just a few years into her thirties, but enough older than me to seem more seasoned and mature. She was unusually tall, and a little forbidding. Actually, a *lot* forbidding: Genuine smiles came easily enough to her, but routine, social smiles—the kind that leave everyone in a room feeling relaxed and happy—did not. On this stern-faced woman and her opinion of my work, my livelihood depended. And now she wanted *my* opinion of something *she'd* written?

Umm, maybe, I ventured, there was just a little trouble with this transition? And this word, here, perhaps it wasn't exactly what she meant?

Elise took back the manuscript and looked at it, hard, the way she always did—no knitted brows, just the blank screen of her face, the outside world absent. For a moment, the room lay still. Until, abruptly: "Oh, yes, certainly." And saying this, she pounced on the manuscript, *pounced*, using her whole body, arms and shoulders, not just her hands, to scribble in the words that made it just the slightest bit better.

Only then did she look up and acknowledge me.

I didn't realize it right away, but that eager, egoless, unguarded "Oh, yes, certainly" *stuck* with me: Thank you, Elise. From a distance of twenty-five years, I write now of a tricky little professional situation. But for her, I am certain, it didn't exist. For her there was no editor or writer, no senior or junior, no man or woman, no vanity, no pettiness, no personalities. There were only the words, and the ideas they expressed, that were our job, together, to get right. Nothing else mattered. And *everything* that mattered was on that page.

I write of a time during the late 1970s and 1980s when I and a few other young writers—freelancers, interns, office assistants, kids just starting out—worked with Elise at the magazine. Most of what I know today about writing, especially writing about science, medicine, and other difficult subjects, I learned then. Others did, too. Those who came to see the ceaseless flow of red ink as the gift that it was went on to great things. They wrote for *Time* and *Discover* and *Life*. They edited the magazines of elite universities. They wrote books, won awards and fellowships, made names for themselves. And their writing lives mostly started in that little office in Whitehead Hall that Elise, with her madcap creativity and breathtaking intelligence—you'll see ample evidence for both in the pages that follow—made entirely hers.

Elise had become editor a couple of years before and had set about making her little bimonthly into something far more than a mere alumni magazine; what the *New Yorker* was to the urbane literary and cultural life of New York City, the *Johns Hopkins Magazine* would be to the scientific, scholarly, and creative world of Johns Hopkins University, with long, thoughtful articles and clear, graceful prose. An anthropologist at work. Cervical cancer. Rockets shot into the sky. An issue following medical students through their four years. Charming little Christmas presents to her readers, like puppets of chimerical creatures. Each year, she and her staff would walk off with awards for fine writing, and twice during her tenure, *Johns Hopkins Magazine* was named best university magazine in the country.

Me? I'd been a freelance writer for a few years, had prematurely tried to write a book, and now, after some time away, had returned to Baltimore, where I was managing the rent on a tiny apartment but not much more. About a year earlier, combing for freelance assignments among local newspaper and magazine editors, I'd made an appointment to meet Ms. Hancock.

It was the perfect time. It was 1976 and Elise was hungry. The university was celebrating the hundred years since its founding, and numerous centennial events—seminars, conferences, and celebrations—were being held. The university magazine, with its two-person staff, was supposed to cover as many of them as possible and needed freelancers to help fill centennial-fat issues. Elise assigned me to attend one of these events, a symposium on decision making, and write about it. I did so capably enough that in coming months she gave me more work.

Capably enough? That didn't mean you were the next Tom Wolfe or John McPhee. Just that you had some slight feel for language and seemed to understand what you were writing about. Elise was always relieved when one of her new writers proved as curious as she was, got the facts right and the story straight. You could have all the word magic in the world, she used to say, but if you were going to misquote distinguished scholars, and skate superficially over the life's work of world-class scientists, and think you were going to get away with spinning pretty verbal webs around what you couldn't be bothered to understand, then how could she work with you? Elise was interested in science and ideas, and she was impatient with writers who weren't. So, while I hadn't the sheer verbal facility of some who came through her door, I had enough of this other quality to keep landing assignments. A conference on war gaming. A peculiar moving-walkway engineering project. Then, longer pieces on recombinant DNA, evolution, the ecology of the Chesapeake Bay, particle physics, laser surgery. Over the next ten years or so, I did about three dozen pieces for her, most of them long and ambitious. And always—at least at the beginning, before word processors, when I still used my old Smith-Corona portable—the time would come when we'd sit down with the manuscript.

This is the part that usually gets freighted with nostalgia, with sepia visions of crisp white paper smacked by those great old typewriter keys, of ink smudges and red editorial squiggles and slashes garlanding the page, and great XXXs smooshing through whole paragraphs. But do you know what those squiggles and slashes and XXXs do? They change your words and ideas, develop them, reorder them, dismember them, turn them inside out, or obliterate them altogether. They signify, at some level, that your literary expression is tedious or crude, your ideas silly, boring, wrong, or off the point. Or that you've left a thought undeveloped or muddled, a scene or story vague, flat, or insipid. Together, they imply that what you've done won't do, and that what the editor has done, through her marks, scrawls, and penned-in changes, is much, much better.

Better, that is, in her opinion. But what if you, the author, begged to disagree?

Well, I did disagree. A lot. Elise's emendations, after all, weren't chemical formulae, right or wrong, but expressions of judgment and taste. And I was too young, sure, and stubborn to accept hers for the wisdom they embodied. So she'd say, This is too much, Rob. And I'd say, No, it's not. She'd say, You need to rethink this, Rob. And I'd say, No, it's fine the way it is. Rob, do you think the reader wants to know all this? Rob, what is it, really, that you want to say?

Most of the time, of course, Elise was right, and I'd later come to see as much. But not without a fight. After all, these were my words—my ideas, mine, me. Every word became a battle, and poor Elise was left to explain why she saw things as she did. Mostly, she did so patiently. Sometimes, though, her normally composed features would tighten into annoyance and her criticisms could be harsh. But one way or the other, sitting beside her at her desk, the manuscript on the sliding desk tray between us, I learned.

I can attest to the wisdom of the writerly injunctions you'll find in these pages because at times I've ignored them all. For example, *Do not confuse a topic with a story idea.* That's just what I did once with a long piece about memory. What about memory? Well, *everything* about memory. Elise helped me save it, almost; I wound up saying that an understanding of memory still eluded researchers, and that it was a multifaceted phenomenon, duh. But the piece was never as good as it should have been because my topic, one of Elise's dreaded noun-ideas, never found its proper focus. It was all over the place. Literally so: The piece was littered with enough sidebars to tell any savvy reader that its author didn't know what his story was about.

Before first meeting Elise, I'd written a mercifully unpublished book about urban life with a good title, *City Sunrise*, but little else of merit. After we'd begun to work together, I let Elise read it. At the time, she was tactful, even gentle. But later, whenever I wrote something that pleased her, any compliments she dispensed would take the form of how, yes, I had certainly made progress since *City Sunrise*.

Even after my work began to enjoy her favor, she'd freely poke fun at its infelicities. My prose, she said, reminded her of a noisy, congested city street, cabs whizzing, pedestrians darting, horns honking, all calling attention to themselves to maddening effect. By now, of course, this image is acid-etched on my brain tissue, helping to pull me back from my worst excesses. And through a hundred such vivid images and stern directives, Elise remains beside me today. She doesn't always win the battle against my writing demons, but she's always there, at my elbow, fighting the good fight against poor form and sloppy thinking.

This, then, is the happy payoff for my pigheadedness all those years ago, one I could scarcely then have imagined: Each time Elise answered my objections or demolished my literary conceits, she'd draw me into the rare and splendid precincts of her mind. And in doing so, she'd bestow just the sorts of insights you'll find in the pages of this book. I speak now not of such matters of common sense and good professional practice as double-checking names, though these count, too. But rather of a rich sensibility of respect. For language. For ideas. For people. For the surprising and the deliciously weird in us all. And most of all, respect for the world, the endlessly enthralling "real" world outside us.

Elise is the supreme nonfictionist; you won't find that word in the dictionary, but I know she would approve. Many writers, unconsciously or not, subscribe to a hierarchy that makes fiction the goal to which any real writer aspires, nonfiction a sad second-best; bitterly they toil in nonfiction vineyards, dreaming of novels and stories they will write some day. Not so Elise. She read fiction, gobs of it, of every kind, from Jane Austen on down, even the occasional romance novel; her imagination was vigorous and playful, enriched by fictional worlds. Yet I never sensed in her any regret at being sadly stuck in a workaday world of real people discovering drearily real things about the immune system, estuarine ecology, or gluons. Rather, I learned from her that there was wonder in the world and that a writer's greatest pleasure was to tell of it.

Tell of it, mind you, not to the already expert but to everyone else. Technical reports for technicians? Scholarly articles for scholars? These had no place in Elise's magazine. When her writers took on stories in anthropology, oncology, or cosmology, they wrote not to specialists or other scientists but to Elise's "educated curious"; this made it "science writing," not "scientific writing" or "technical writing." Science writing is so hard to do well because it dares aim intellectually formidable material at just those readers presumed to have little background, education, or interest in it.

Science writers and editors needn't start off knowing much science. Some of the best of them do, but some of the best of them don't. They must, though, be able to learn science, be eager to wade into its complexities, ask intelligent questions, and shake off the high intimidation quotient of a dense, jargon-laden article in the Proceedings of the National Academy of Sciences. Elise was a member of this breed; she was an English major in college and took only a handful of science courses. Yet in sending her magazine out to joust with science, medicine, and technology, she was fearless.

Once, long ago—before the genome project, before the rise of the big biotech companies—two Johns Hopkins researchers figured out how to snip DNA, the molecule that embodies life's genetic heritage, at particular points. Pretty soon, scientists were taking pieces of DNA from bacteria and slipping them into other organisms. Some people began to worry about the dangers and called a meeting at the Asilomar conference center in California to discuss them.

This was a big science story and, since Hopkins researchers had played so crucial a role, a big Hopkins story. Elise resolved that Johns Hopkins Magazine would cover it—more particularly, that I would cover it. What a team! She had no grounding in molecular biology. I had never taken so much as an undergraduate biology course. But so what? We could do it. And we did. The result was "Pandora's Box, Chapter XI: Splicing the Double Helix." It reads a bit breathlessly today. But, then again, that was the atmosphere of the time, even among some normally circumspect scientists. And our readers became conversant with issues that, in new forms, linger with us today.

I learned a little biology. But more, much more, I learned to swim out from shore and into the rough seas of hard science, and not worry too much that I would drown.

Over the years, I've kept a journal of writing advice that I share with my students or otherwise draw from. I'd thought of this as altogether fresh, reflecting my own experiences as a writer, my own particular take on things. So it was chastening to read Elise's book and see that many ideas and insights I'd thought were distinctively mine were, in fact, distinctively hers.

Oh, at times I found myself thinking, No, that's not how I'd do it. Elise says to use a tape recorder, that all journalists do. Well, most journalists do, but not all; I don't. Elise says that after immersing yourself in your material you should hold off writing, think things through first; begin writing only "when you're clear enough that you won't go wrong." I never get that clear. I use the act of writing itself to find that elusive clarity, slogging through swamps of nonsense and incoherence to get there.

But far more typically, there they are in black and white—insights, ideas, strategies, and preoccupations I'd identified as mine plainly culled from Elise over our years together. For example, her highlighted, boldfaced kernel of pure Elise wisdom: LISTEN, really listen. And mundane things: How, before an interview, you carefully set out written questions—then, during the interview, mostly ignore them. How, before you write, you compose a headline or title: I'd always been pleased with myself for abandoning the more common advice that a title came last, after the hard work of writing. Yet here Elise reveals it for the profound compositional trick it really is. Writing a headline, she writes, "will force you to get precise about both topic and approach."

Reading Elise's book today reminds me that wisdom and good sense can get passed down, that sometimes we truly can learn from one another.

I am so proud to be Elise's student. Read this book and I suspect you will be, too.

Robert Kanigel

Acknowledgments

Many thanks to John Marcham, former editor of the *Cornell Alumni News*, who trained me; to E. B. White and Will Strunk Jr., who were my mentors through their classic book; to Rob Kanigel and Jackie Wehmueller, who egged me on; to Jack Goellner, Katherine Hancock, and Mary Ruth Yoe, who showed the manuscript no mercy; to the many patient scientists who coached me on their work; to my friends and colleagues at other university magazines, who were always just a phone call away; to the dozens of writers with whom I sat on the floor and argued over manuscripts until we got it right; and to the readers of this book, who I hope will carry on the good work with the care and integrity I have valued in all these people.

Ideas into Words

A Matter of Attitude

Only connect.
—E. M. Forster,
Howards End

To write nonfiction, whether "science writing" or any
other kind, is an act of intimacy. You are inviting the
reader into your world—into your mind, no less. As your
close companion, the reader will share the alien cadences
of your thought. He will borrow your vocabulary, no
doubt of a flavor not quite his own. He will be at the
mercy of your skills to see, to hear, to think and feel, to
assess people and draw them out, to persist until you really know—and, of course, to put what you know into
words. It requires a certain trust, to be a reader.

Once the words are in print, however, it's the writer
who has to trust, because the reader now holds the reins.
If an author loses me, I can stop reading. Or I can skip a
chapter, or three, or skim, or read each paragraph five
times, analyzing and underlining in several different colors until the words droop and die. Whatever the reader
does, the writer has no recourse.

Yet how intertwined we are, reader and writer, sharing
a universe of words. Reading, I can sit with Loren Eiseley
in doleful twilight and ponder a skull. I can hitch a ride as
the mind of Stephen Hawking soars through all of time
and space. Diane Ackerman gets me drunk on the sensual
beauty of planet Earth and its creatures, while Sebastian
Junger propels me into danger, forest fires and storms at
sea, and tells me how they work. A self-help book may
bring hope and guidance as much as or more than information. For the moment, reading, I am not alone.

These effects are not accidental or random. Good writers always have the reader in mind, not only as they write
but also in the finding out that comes before. They do
their research with integrity, digging deep, and they write
with the same care. They connect as deeply with the ma-

terial as they do with the readers. Indeed, their curiosity and its fruits are a large part of what the reader senses, of what lets the reader trust them—a process that begins with the first sentence.

When I write or edit, particularly as a piece opens, I literally feel myself to be reaching out to someone. It is as if I tap a hundred shoulders. Look, I say. Look at this, see what I found. Isn't that something? . . . And we walk forward, reader and writer, and explore the world together. When I read, it's the same transaction. As I start reading, I am meeting a person, and I am deciding, in just about the millisecond it takes in real life, whether I want to talk with this guy. Does he know anything I want or need to know? Is it comfortable breathing the same air? Can I trust him to get it right? And will he promise not to bore or puzzle me in the meanwhile?

If you have never sat in a train station and watched someone flip through a magazine, try it some time. It's humbling. About a third of people flip from the back, not the front (which is why many magazines run those inviting final pages of essay or photo), and the pages turn about once a second. Flip . . . flip . . . flip . . . two-second pause; no, not this . . . flip . . . flip . . . flip . . . three-second pause; eyes are scanning . . . flip . . . flip . . . till finally something catches. (As the reading begins, there is often a small, overall shake, like a bird settling onto a nest.) Keep watching, though—the reader may quit several paragraphs in, if the initial promise is not fulfilled.

Since you are reading this book, I am assuming you want to be that writer, the one who catches the reader, then delivers the goods. You want to be a person who can find something worth sharing and capture it in words. As for me, I want to help you become that person—both to BE that person and to DO the work.

A lot of the Doing is skill; to have any useful inklings about people, communities, science, or the natural world is a large skill, and so is writing. You will need both abilities, preferably based in good brains, education, and talent for making the language sing. (But hard work helps more than one would think.)

Beyond that Doing (and possibly the hardest part), you will need to Be the sort of person whom readers trust with their attention—and the readers cannot be fooled, because they have crawled into your mind. If you are bored, the reader will be bored. If you are skating on thin ice, unsure of the information, readers become uneasy. If you are counting on a first draft to be good enough, the reader will flip on by. Worst of all, readers can tell when you're showing off and unconcerned with them. They don't necessarily make the diagnosis, but they do feel annoyed and ... flip.

It is a lovely moment, there in the train station, to watch someone absorbed in an article that you have written. Few people write enough to see it often, but it happens and you will like it. Fine—but that moment is a bonus. If praise is your purpose, your writing will misfire. People know. The words and organization of your writing have to grow out of the material, which must authentically intrigue you. And not only do you have to care, you have to care so much that you can hardly keep it to yourself. At times, you may feel like a kindergartner rushing home to tell your mother the great news: Red paint plus blue paint makes purple! The emotion can be that intense. You must want to know—generically, about everything—and you must also want to share it. That's the Being part of life as a writer, and also what this chapter is about: a series of ideas, attitudes, and habits that will help you become that person.

Once the right attitudes are in place, the Doing gets easier, because it has roots, and it will occupy your attention so fully that you get out of your own way. You won't be squandering thought on yourself and how well you are writing. Instead, you'll be fixing one mental eye on the reader and the other on the fascinating thing you have found, and you will write by laying out the details that make up your mental picture. Basically, you'll just be talking to the reader, as to any other person in your life, except this talking will be in writing. As a process, writing as if you're talking is easier (and more effective) than manipulating technique. Not only that, all the techniques make far more sense when they are grounded in the social skills you've been practicing all your life: connecting with other people.

The first step to writing nonfiction, especially science, is to know that you can do it.

Do not let new material intimidate you: it's okay to be a beginner. The moment you believe that you cannot understand something, whether it be a physical science, a social science, or the Dead Sea Scrolls, it will be true, so don't admit the thought. Just don't go there. Instead, tell yourself, "I am a beginner at [whatever it might be]." Grant yourself a learner's permit.

That thought is so important I'll say it again—*grant yourself a learner's permit*. Enjoy your ignorance. It's exciting. Every time you tackle a new subject, you are doing something that will take you into a new and bigger world.

In fact, within reasonable limits, ignorance is an asset. It is likely that you will never understand the world in the way a scientist does—but the readers don't either. When you ask "stupid" questions, you are only asking what the readers would ask if they could. Because you do not know, you will nose out the gee-whiz examples and unspoken assumptions that the scientist is apt to take for granted. ("Huh? Everybody knows that.") No, everybody does not. And you, beginning with a learner's permit, will have a good sense of how much to explain and how much to gently sidestep.

When you are a learner, it is okay to grind the gears and drive slowly around the high school parking lot. In fact, you are not only allowed but *expected* to be slow and clumsy for a bit, both in your writing and in your understanding. Then, as you venture out onto the road, believe me, the researchers will be delighted to coach you.

Over the course of a career, you'll need hundreds of learner's permits, so you might as well enjoy the process. Plunge in with a good heart. If you look up all the important basics and keep asking questions, I promise you can grasp the central concepts. Then, remembering what used to puzzle you, you can design an explanation that the public can understand—including senators, CEOs, religious leaders, and heads of state, all the people who will determine the future of our world.

Just think: As a science writer, you will have a license to go find something new and interesting about how the world works, and then another something, and another, and another. For the rest of your working life, you will get paid to talk to people and pass along the great stuff you find—which can make a difference in the world. I am very proud of the work I and others did explaining AIDS to the world back in 1982, when a panicky public still feared you could catch HIV from a toilet seat.

Go explore. It's a big world out there, and never have human beings had a greater need to understand how our planet and our own human nature work. As a writer, you can make a difference.

Do not let writing intimidate you; you already know how to do it. Or at least, you know the hard part—how to catch and hold someone else's attention. You have been practicing that art since you were two months old, and by now you're pretty good. You're bound to be. When you write, transfer in your social skills. Think of "the reader" as if you were speaking directly, or perhaps writing a letter, and plunge right in. Just write down whatever you would say. In that way, you will automatically avoid the big mistakes—starting with the punch line, for example, or droning on and on about how fascinating this stuff is going to be, or explaining all the concepts that the one you are about to explain must not be confused with. Instead, you will intuitively go straight to the heart of the story, just as if you were telling a joke, and with similar good results.

I mean this advice literally. Whenever you get stuck while writing, stop struggling. Close your eyes, visualize a specific, living, breathing reader, and say to yourself, "What am I really trying to say?" Whatever the answer, write it down. Polish later if it's needed—but you may be surprised at how trivial the polishing can be.

The same approach will help you before the writing, too, while you decide where the story should focus. Again, imagine the reader—your Aunt Gayle, perhaps, or any other person you know who can stand in for the expected reader. If you were going to have lunch with Aunt Gayle, what would you want to regale her with? Of course—there's the story idea.

We'll talk more about addressing your reader in chapter 4, on the process of writing.

Do not let scientists intimidate you; you will find them quite congenial. In my observation, the temperaments are similar. Both scientists and science writers are bright, curious, intuitive, analytic, unafraid of something new, and dogged once they get their teeth into a puzzle. Both like to think. Science writers, however, also need to play with language, and they have a curiosity too big and too restless to stick to one field. In fact, until they find journalism, these people tend to look "undirected," "unfocused," or even *nosy*.

They simply feel they have to know—as do scientists, in a more focused way.

This affinity is one reason that I believe you will find it easy to work with scientists. The other is that the two groups share a goal, helping the public know what science does and why it matters. When you approach scientists in a spirit of collaboration, and when you have done your homework and show every sign of respecting their time and knowledge, all will go well.

And in the rare case of trouble ... frankly, I have noticed that the higher a person's scientific reputation, the less interest she has in rewriting quotes or insisting that you show her the copy. The big guys have better things to do. Maybe that's how they achieve so much.

Later in the book, especially in chapter 3, on interviewing, we'll talk about the specifics of how best to work with scientists.

Stay in learning mode. I apologize for the cliché, but it's true: You must be willing to learn, as a matter of attitude. Without hard work, great gifts of the mind, eye, and spirit will come to nothing. Conversely, if you have even a small gift for words and ideas, you can eventually do very well simply by keeping at it.

Fortunately, when you have found the right field for you, "work" feels more like fun than like labor. Think of anything you do for fun—a sport, cooking, macramé, competitive bridge, caving, anything—don't you enjoy reading and learning about it? You'll find the work of writing science to be the same: not always easy, but enjoyable. Even the chore parts of it, like keeping up with current research, will have a certain zing, like panning for gold.

We learn at all times, not only when we plan to. Therefore, hold your junk reading to a minimum—junk meaning anything you do not want your own thinking and writing to echo, because echo it will. That is why years of teaching undergraduates can be so deadening to a writer: it steeps the mind in sophomoric prose. Go thou and do otherwise. Steep in the great and good of any type: fiction, essays, history, science writing, whatever.

Compared with fiction or poetry, science writing relies far more on content and clarity, far less on lyric intensity. That's a matter of emphasis, however; the components of good writing remain the same in every field. It follows that you can learn by doing and reading *any* kind of writing—so long as it's good. Poetry will sharpen your sense of the power of a single word. Fiction will show you a trove of technique, for the human mind loves narrative. What happened next? What happened next? We all want to know, because the human mind is built that way. The more you learn to tell stories, the more your readers will love reading what you write.

When you work with a teacher or editor, nix on thoughts like "It's only grammar" or "I'm right" or "I've said what I want to say and that's it." No matter how good you are, you can always get a little better. When people try to help you, listen. They may be onto something—or not, but you will never know which if you do not listen.

Once, at an editorial conference, I walked out of a room behind two people who were grousing about the talk just delivered by John Bethell, for many years editor of the prize-winning *Harvard Magazine*. "I could do a great magazine too if I had his budget," said one. "Yeah," said the other. "Fat chance. He's really lucky . . . See you at the bar?" Ten minutes later, while the grousers were presumably at the bar, I walked by the room of exchange copies—and there was John Bethell, combing through other people's magazines to see what he could learn.

Make no effort to be original. As adolescents, most people try to pose as someone they are not—at least my friends and I did. Alas, posturing never worked as well as being ourselves.

Trying to "be original" in your writing is much the same. It's almost universal among the young, and it's a waste of effort, because you already *are* original. There is no one else who sees the world or uses the language precisely as you do. Nor has anyone else done precisely the interviews and research and thinking you have done. So relax. Save all your energy for understanding the subject, and as you write, keep asking yourself, "What am I really trying to say?" Then say it. The result will be original.

Not only that: When you can reliably know what you want to say and say it, you will have discovered your mature writing voice. It's as simple—and hard—as that.

Science writing is seldom self-revealing in the way that poetry is. At the same time, to write *anything* is to expose yourself, unavoidably. You cannot help but reveal the way your mind moves, whether it's quick hops or a delightful ramble or an inexorable drive. People will see whether you gravitate to the hopeful aspect of a subject or the Big Moral Issue; they will sense your attitude toward the reader (probably much like your attitude toward people in general). All that shows, and more. Indeed, that intimate connection with you, mind to mind and spirit to spirit, is part of what readers seek.

As I write this book, I have been a professional writer for more than thirty years, and I still have moments when I think, "I'm going to tell them *that?*" ... Unhappily, it is the very thoughts we fear to reveal that are, in fact, original.

Rough drafts are by definition rough. Therefore, invest your vanity in the finished piece of writing, not in your rough drafts. Once you think that way, despite having a vanity of normal size, you will be able to listen to would-be helpers without feeling under attack. You can think, "Well, of course there are problems—I'm not finished yet! It's only a draft!"

Aren't those comforting words? "It is only a draft."

"It is only a draft" will also set you free to experiment, as well as to persist. "It is only a draft," you can say to your inner critic. "I'm just getting it out on paper so I can see what I have. I'll fix it later. *This is only a draft.*"

Please notice that I am not urging you to "have confidence" or "keep up your self-esteem." Several times when I have failed as a teacher, the pupil has had very high self-esteem—so high that nothing I said could be heard. Rather, I am urging you to realize that your writing improves as you work on it. That's a fact you can count on, and it's more useful than self-esteem because it will lead to constructive action. Whether you diagnose a glitch or someone else does, you can be thankful: "Great! Now I see what to do! Luckily, it was only a draft."

Acquiring this attitude may take a little practice. For some, it may feel phony the first few thousand times you try it on. If you keep self-talking that particular self-talk, however, you will become a writer with whom editors want to work and whose work keeps reaching new heights. By a Zen-like paradox, subordinating your own needs to those of your reader and your material will help you go far indeed.

**Moments of frustration and despair can be a good sign;
they are an expected part of learning,** so much so that animal trainers count on them. They call them "prelearning temper tantrums," because the frustration means that the creature is about to get it.

Neurologists have a saying: "Neurons that fire together wire together," and they mean that literally. When a group of neurons fires hard enough to activate other neurons, the receiving neurons actually create new receptors to hold the connection, which is called a "neural pathway."

To illustrate the point, a neurologist once picked up a small, black rock from his desk. "Catch," he said, and as I caught it, my hand flew shoulder high. The "rock" was not heavy, as expected, but featherlight, a piece of foam. When we did the toss again, however, I could not duplicate the motion. Once I knew the object was foam, I could not help but catch it lightly. In a single catch, a neural pathway had formed—and that's the way animals learn, including human ones.

The harder the firing, the stronger the pathway. That is why memories from combat or other trauma can remain so vivid and trip so easily.

The more a pathway gets used, the stronger the connection. We strengthen neural connections whenever we practice, whether it's writing, a tennis serve, or fuming at our in-laws.

Frequency also matters, which explains why it is better to practice new skills for fifteen minutes daily than for three hours just before the lesson.

Coming back to you and me struggling to write: Sometimes the struggle is with the writing itself, sometimes with the topic, sometimes both. In all cases, it's a big learning, because not only must we create many, many new pathways, but also they must be complexly interlinked. So it makes sense that we might need an emotional boost, namely that big adrenalized surge of irritation, to help us fire enough neurons, all at one time, to create the new network.

What to do? Mainly, draw strength from the fact that you're ready to scream. It means you are nearly there.

Be a writer at all times, not only when you sit down at the keyboard. The more you live as a writer, the easier it will be to write, because much of daily life will serve as practice. The way you speak, listen, watch, and read will have much to do with how well you write.

Speak precisely. Make it a habit to say what you mean, rather than settling for a close approximation. If you're not sure what you mean, say something like, "I need a minute to think about that," so that other people know you're not ignoring them. As a secondary benefit, you may blossom into a wit, because once you routinely capture a scene in three or four words, people will find you hilarious.

Practice composing descriptions and narrative in odd moments when you might otherwise be bored, as when waiting for a train or a friend. Wherever you are, look around and ask yourself how you would write about what your senses report. Familiar objects and people, especially clichés (a beautiful sunset) can be surprisingly hard to capture, while human activities and interactions are even more so. This practice carries two huge benefits: not only are you practicing narrative, but you are also developing the art of seeing freshly. A useful question: If I were a tourist from another place or time, what would I notice here?

Practice separating phenomenon from conclusion, in particular about people. When you notice yourself saying or thinking some judgment—"She was really angry," let us say—ask yourself what you saw, heard, or felt that made you think so. This practice is so demanding that I don't think I will *ever*, fully, know my conclusions from my data. (A Buddhist would say that, if I did, I would achieve satori and dematerialize.) We can aim to know, however, a practice that has two benefits. In your personal life, it prevents a lot of unnecessary pain, because it helps you spot the times when your conclusion comes from your history, not the here and now. (This woman did not say or do anything to make me think she was angry—I just *expected* anger. Oh.) In the working world, the distinction keeps your observation circumstantial. Compare "Dr. Jones was angry" to "Dr. Jones stiffened and her mouth became a thin, pale line." Which version is more compelling and believable?

When you must sit through a forgettable speech, practice editorial pruning. Rewrite as you listen. (I sometimes used to "type" the leaner version with my toes inside my shoes, an invisible form of fidgeting.) In this way your face will look attentive (because you *are* attending), and you might even remember what was said. If the speech resists pruning, wake up. You are hearing a masterpiece.

Build a library, both of stuff that is just good reading and of background material, such that you can go from total ignorance to basic grasp overnight. The Britannica on-line is a great resource, as are the various fifteen- to twenty-pound tomes in which some authority spells out Everything We Currently Know In Our Field. If you buy such a tome from the remainder table, some few pages will not represent the latest knowledge, but no matter: You are seeking background, not the latest—*that* you will get from journals and researchers.

When you read, read as a writer. That is, pay attention to your own reactions. When you find yourself enthralled, pause to ask yourself why. (What reached me? Where did this effect start to build? Is this a technique I can use?) Conversely, if your attention wanders, see if you can isolate where and how the writer lost you. Was there simply too much material? Something about the tone? Was it too vague? Oversimple? Too dense? Too abstract? Did you lose hope that a point would emerge?

Everything is interesting. Take this idea as an item of faith. If a subject bores you, it's because you don't know enough or you've adopted the wrong vantage point. Keep poking till you see a sign of life. Ask people who do care to tell you why.

I once took a ballet class in which the teacher insisted that, if our legs felt like jelly, we should jump harder. To my surprise, I found he was correct. Could it be that the increased effort elicits a rush of adrenaline? You will find that the same is true of life in general, including the finding and writing of story ideas. If you feel bored and your mind flobs around like jelly, pay more attention, not less.

Incidentally, the whole subject of attention in psychology is a hot one. Paying attention appears to organize the brain; it makes possible both learning and memory.

Feel free to use weird words or even to make words up, an idea of which I was reminded by the word "flob" in the last item. Did you know what I meant by flob? I am sure you did. Is the word in the current Random House dictionary? No, it is not. Is there a better word for that meaning that does already exist? Possibly so, but I don't know it.

"Flob" does appear in the Oxford English Dictionary, however, as well as in my aged unabridged Webster, which defines it as "to be clumsy in motion" and calls it rare. Too bad—I thought I had made it up. Instead, I must have remembered it from *The Thirteen Clocks* by James Thurber, a book that also features a creature called the Todal. The Todal is "a blob of glup," an agent of the devil sent to punish evildoers for not having done as much evil as they should, and when it comes for an evildoer, it *gleeps*.

I see no reason why we and our readers should enjoy our language less than Thurber and his readers, so I offer you "flob," with which to describe the movement of Slinkies, sloths, or stranded jellyfish, and thousands of other words just as great. Seriously. Rich with onomatopoeia, recombinations, and words lent by other languages, the English language has a range and flexibility like no other. Feel free to exploit this glory of our tongue, so long as your meaning is clear. Our readers deserve all the lively writing we can muster.

The reader is smart. Research has repeatedly shown that high intelligence correlates poorly with advanced education or high socioeconomic status. Rather, it is scattered through the population. Not everyone would agree, but I hope you will, because if you believe it, you will never condescend to your readers.

Do not dumb down your train of thought, whether you are writing for the *New Yorker* or your local hospital's newsletter. While less educated people may need to be addressed in basic language and universal images, they are fully capable of understanding the issues, and they will detect condescension in a flash.

Old-fashioned teachers used to make kids write things out one thousand times, as both punishment and lesson. ("One keeps appointments. One keeps appointments. One keeps appointments. . . . ") I know you won't write it out, but it will not hurt you to say this maxim several times: "The reader is smart. The reader is smart. The reader is smart. . . . "

"Fifty percent of everything I am teaching you is wrong. The trouble is, I don't know which 50 percent." The great British physician Sir William Osler said that a century ago, in his famous teaching rounds at the fledgling Johns Hopkins Hospital, and the point still holds true. Just take it as a fact: Some to much of today's research is mistaken or incomplete. Furthermore, the longer you live, the more of the "facts" in your head will be erroneous. Science keeps moving and the Todal takes the hindmost, so make sure you stay current.

Beyond purely scientific assessment, you should develop a well-honed skepticism, holding two questions in mind:

Who funded the research? Industrial or politically driven funding does not automatically invalidate the work. For example, I have no doubt that a daily glass of red wine does benefit cardiac patients, especially since grape juice and raisins have the same effect (or so I have read in *Science News*). Still, the reader is entitled to know that the original research was industry funded. Stay awake, especially for blandly named "research institutes" with no academic affiliation; such are often nests of lobbyists. If you use material from such a source, do the reader a favor: name the funder and state its nature—in some suitably neutral language, of course, like "conservative think tank." (I just said "nest of lobbyists" among us writers.)

It seems unfair to single out private funders, however, especially as what the federal agencies choose to fund is also of public interest. So I suggest that we add funding to the famous five Ws of good reporting—Who, What, Where, Why, When, and Wherewithal? Just tuck the funding in a subclause.

The second question is harder to answer:

Does this research fit the values and preconceptions of our era just a tad too well? Do we really want it to be true? The problem here is that we live in our own time and place as fish swim in water—we only get the view that's visible from where we are. Isn't the whole world water? Fish probably think it is, and we likewise think ... hmm. What *do* we think? What are the limitations of our current ways of thinking?

Take artistic forgeries, for example. When Victorians looked at contemporary forgeries of ancient Greek vases, they saw authentic perfection, because each one matched their particular ideas of classical art. When we look at those same vases today, we say, "Oh, how Victorian!" I would love to know what people will see a hundred years from now when they look at the successful forgeries of 2002 and say, "Oh, how Bushian!"

The blinding preconception is most obvious in the social and medical sciences but can be found in every field, including the data-heavy ones like physics. You might try asking researchers about it. What do they think might be blinding researchers in their field right now? The answers are never dull. And you should think about it yourself. You will be fooled, but less so if you stay awake.

For example, I am riveted by all research about complex interactions in the life sciences and skeptical about all research that seeks "the" cause or "the" remedy or "the" active ingredient for anything whatsoever. Western thought has a bias toward single actors, perhaps because germ theory and antibiotics succeeded so well—for a while.

What do you think?

Scientific truth is not a matter of opinion, not even in the so-called "soft" sciences, like sociology or psychology. Scientific truth is a matter of evidence. That is the attitude of scientists and one you should cultivate consciously. You'll need to work at it because our cultural undertow tugs the other way: Americans tend to think pro versus con, two sides to every question—like lawyers. That bias makes sense, as the United States has so many lawyers per capita, but it equips us poorly to look at scientific evidence.

In the legal model, "evidence" is opinion, something each side can purchase in a flavor that supports its case, while "truth" is either irrelevant or does not exist. Lawyers have often told me so, seeming shocked at the very word. ("Truth?" The eyebrow rises.) In law there is no truth but only verdict, which arises from two equal positions duking it out. And that's okay. Five hundred years from now, it will make no difference whether Sacco, Vanzetti, and O. J. Simpson were "truly" guilty.

An unconscious habit of legal thought can betray us when it comes to science, however. I suspect it misleads many nonscientists to eavesdrop on scientific debates as if "evidence" were a ploy, then to make decisions as if none of it will matter five hundred years from now.

Scientific evidence is the result of experiment or careful observation, based on rules devised to make sure that, if other people do the same work, they will get the same results. In other words, evidence is supposed to be true.

A good scientific theory wins by best approximating reality. It must account for whole swathes of evidence, preferably from many directions, and it must be able to answer any sensible objections—not just shoot them down with a wisecrack or persuade a dozen key decision makers, but *answer* them.

As you write about research, it will be important to stay

with the evidence, in a scientific sense. Evidence will not only hold your train of thought on line but also help the public think straight. In the next few years, the people of this planet will be making choices that determine whether our great-grandchildren can live here, and the way you and I write about science will have a lot to do with what happens. Will we and our leaders know to set policy according to evidence, however unpleasant?

A paradox: Because research focuses forward, into the new and uncertain, scientists tend to talk about "truth" as if it were the Holy Grail—always sought, often glimpsed, never within reach. Scientific truth is always partial. Yet, for the most part, the best scientists agree on the current best theory, which they recognize because, well, it fits. It answers the most questions with the greatest precision and the fewest loose ends. There's a satisfactory click, as in some arcane gadget from the old Abercrombie & Fitch, and everything comes together. Aha! Here's bedrock, a firm place to stand. And then, from that bedrock, one sees a new world. (Well! If that's true, then ...)

Think of the scientific process as if you faced a mountain of jigsaw pieces from several puzzles—or so you think. But you cannot be sure, because pieces may be missing. Nevertheless, you start sorting your evidence. These two are not only the same ineffable blue, but innie and outie link precisely. These two ... no, here's another one ... these three look like they might be parts of a basket but do not link. And here are some leaves, and several pieces with a straight edge, so they're from a border. And so on. You have a hunch the ineffable blue might belong with the basket, because the scale is identical, but you won't know till you find a connecting piece, one that flashes both blue and basket. Or could the blue and basket pieces be part of two similar puzzles? ...

When you hear scientists explaining their work, you will often sense that kind of process. For example, listen to the late lock-picking, bongo-loving, Nobel Prize–winning physicist Richard Feynman, here in a passage from Surely You're Joking, Mr. Feynman (Bantam, 1986). He is in an ecstasy of discovery because he has just seen the solution to beta decay. (Don't worry about what beta decay means, which is here beside the point. The point is to feel how a scientific puzzle clicks together, so that you'll recognize that click in the future.)

The passage begins as three of Feynman's colleagues at the
California Institute of Technology tell him that the evidence

"is so mixed up that even some of the things they've estab-
lished for *years* are being questioned—such as the beta
decay of the neutron is S and T. It's so messed up. Murray
[Gell-Mann] says it might even be V and A."
   I jump up from the stool and say, "Then I understand
EVVVVVERYTHING!"
   They thought I was joking. But the thing that I had trou-
ble with ... —the neutron and proton disintegration:
everything fit *but* that, and if it was V and A instead of S and
T, *that* would fit too. Therefore I had the whole theory!

Did you see it go by? He's just chucked out a piece that belongs in some other puzzle, if anywhere. In a flash, he sees
how the picture *has* to be. And that night he crouches at the
kitchen table till dawn, working out his new equation. The
sky and basket join, and the leaves turn out to fill the basket.
The innies and outies fairly fly together.

I calculated all kinds of things with this theory. The first
thing I calculated was the rate of disintegration of the
muon and the neutron. They should be connected together, if this theory was right, by a certain relationship,
and it was right to 9 percent. That's pretty close, 9 percent.
It should have been more perfect than that, but it was close
enough.
   I went on and checked some other things, which fit, and
new things fit, new things fit, and I was very excited ... I
kept on calculating, and things that fit kept on tumbling
out: they fit automatically, without a strain. I had begun to
forget about the 9 percent by now, because everything else
was coming out right.

Then something knocks on the window, TAC-TAC-TAC, and he
sees a white face looming against the night. He screams! ...
But that's another story, vintage Feynman. The white face
was a disgruntled would-be girlfriend. (As Feynman tells his
stories, he always clings to the opposite pole from
pompous.) Anyway, in the morning, he tells his colleagues
that he's got the theory all worked out.

"Everything fits."

Christy, who was there too, said, "What beta-decay constant did you use?"

"The one from So-and-So's book."

"But that's been found out to be wrong. Recent measurements have shown it's off by 7 percent."

Then I remember the 9 percent. It was like a prediction for me: I went home and got this theory that says the neutron decay should be off by 9 percent, and they tell me the next morning that, as a matter of fact, it's 7 percent changed. But is it changed from 9 to 16, which is bad, or from 9 to 2, which is good? . . .

I was so excited that I couldn't think. It's like when you're rushing for an airplane, and you don't know whether you're late or not, and you just can't make it, when somebody says, "It's daylight saving time!" Yes, but *which way?* You can't think in the excitement.

You can guess, of course, that the shift was 7 percent to the good. Everything fit.

If you have not read this book, you should run right out and get it. And . . . you would know the click if you heard it again, would you not? If so, even as a nonscientist, you can now decide which basic science is hot stuff. You will know what stories to pursue by that click, plus the ensuing excitement. It's a one-two punch.

When scientists say they are "unsure," ask into it. Sometimes they would bet their newborn child and can tell you why. (They'd like more evidence, but everything fits, including a few surprises, factors no one had previously known were involved.) Yet they say "unsure."

The more critical the issue—I've seen this lag with AIDS, climate warming, and infectious disease—the longer scientists feel obliged to say "unsure." In fact, they insist on "unsure," *even when they are sure* (in the ordinary sense), because their idea of truth is so lofty. Also, they feel responsible not to scare the public.

I well remember from the early 1980s not one but several physicians carefully explaining that they did not know how many of the people infected with HIV would become ill. "So far, what we know for sure is that 30 percent do." They would go no further—"30 percent; we cannot say more than 30 percent"—till they were sure I had the message. Then they would shut the office door. "But let me tell you what I *really* think! I think it's going to be all of them, or almost all. This is going to be *awful!*—an old-fashioned epidemic like none of us has ever seen!" The better the scientist, the larger the scruple and the more he insisted on that 30 percent.

So when a scientist says unsure, ask into it, though not for attribution or any immediate use. If you promise discretion, you must be discreet. But the more you know, the better you can plan future stories and get your emphasis right.

It also matters *who* is unsure. For example, Linus Pauling, another lively Nobelist, confounded his scientific colleagues by taking up the cause of vitamin C in the 1970s, at a time when vitamin supplements were in low repute. If you asked most physicians about vitamin therapies, they would smile politely and say that a well-rounded basic diet was all that anyone needed. Yes, nutritional needs were the same in everyone, sick or well.

Linus Pauling, however, had jumped from chemistry into medicine and was promoting vitamin C for staving off colds. "It sounds like nonsense," one biochemist told me, "except for what we call the Pauling effect."

"The Pauling effect?"

"Yes. If Linus Pauling thinks so, you'd better think again. That's the Pauling effect. He's almost always right."

Even at less exalted levels, there are certain scientists whom the others all respect, because they bring a certain breadth, a largeness of vision, yet also think precisely. They ask good questions. They know the latest. They never eat lunch alone. It will pay you to find out who these gatekeepers are and befriend them, because you, too, need their wise judgment.

So-called scientific "mavericks" are generally to be avoided—not always, but generally. In a profession that makes a fetish of evidence and preserving an open mind, the true scientific outsider, fighting the same lonely battle year after year and claiming that no one will let him publish, is unlikely to have a case. Research opposing the current model can and does get published.

Until the early 1980s, for example, "everyone" "knew"

that ulcers are caused by stress, diet, smoking, and so on.
Any other idea seemed outlandish. Nevertheless, when a
medical resident in Australia produced evidence linking ulcers to *Helicobacter pylori*, a spiral bacillus that infects the stomach lining, he was able to speak at a conference and be published in *Lancet*, a British medical journal of high prestige.
The idea came on slowly, but it came on. Today, specialists
generally agree that most stomach ulcers are caused by this
infection, which responds to antibiotics.

Likewise, virologist Peter Duesberg's research attempting to
show that the HIV virus does not itself cause AIDS has been
published in peer-reviewed journals, even though ridiculed.

Use your common sense, in aid of which these guidelines
may help.

Scientific Delusionary	Scientific Visionary
• very much alone	• has current credentials and
• feels persecuted	affiliation; indeed, may be
• starts from a premise,	quite young, just starting
some particular thing that	out as a researcher
is true or untrue, and pro-	• tends to say "a few of us"
ceeds by "what-ifs" and	at various institutions are
"no one can explain"	pursuing the controversial
• has been published pro-	line of work; will refer you
fessionally somewhere be-	to those others
tween never and years ago	• feels excited
• footnotes articles from	• starts from data, then asks
popular magazines as	a novel question or adds
opposed to professional	new evidence, often from a
journals	different field (e.g., may
• seeks publicity	bring what chemists know
• would strike you as a nut	to biology or vice versa)
if met in some other	• avoids publicity
context	

**Practice knowing whether you know or whether you sort
of know.** Daily life gives you dozens of opportunities to
experience both mental sensations—pay attention.

In daily life, we "sort of" know a lot: addresses, for instance. I don't really know the addresses of my friends. What
I know is how to get there and what the place looks like,
which would not help me write so much as a postcard. For
example:

Jack Goellner and Barbara Lamb
End-of-Row with the Beautiful Garden
Left Side of That Curvy Road just before the Calvert School

To write about something, you need to *know* it. You'll be several steps ahead if you can routinely know which kind of knowledge you have on any given subject, writable or sort-of.

Cultivate your curiosity. Writing science is one way to explore how the world works. Why are people the way they are? And societies, and raindrops, and galaxies, and stem cells? Inquiring minds want to know—or at least mine does. Curiosity is a major asset, both professionally and personally.

To show this trait at work, here again is the irrepressible Feynman, back when he was a graduate student at Princeton. He has struck up a conversation with a house painter in a restaurant:

> The guy seemed to know what he was doing, and I was sitting there, hanging on his words, when he said, "And you have to know about colors—how to get different colors when you mix the paint. For example, what colors would you mix to get yellow?"
>
> I didn't know how to get yellow by mixing paints. If it's *light*, you mix green and red, but I knew he was talking paints. So I said, "I don't know how you get yellow without using yellow."
>
> "Well," he said, "if you mix red and white, you'll get yellow."
>
> "Are you sure you don't mean *pink*?"
>
> "No," he said, "you'll get yellow"—and I believed that he got yellow, because he was a professional painter . . .
>
> At this juncture I was thinking, "Something is crazy. I know enough about paints to know you won't get yellow, but he must know that you *do* get yellow, and therefore something interesting happens. I've got to see what it is!"

So Feynman went to the five-and-ten and bought white paint and red paint and brought his purchases back to the restaurant.

> I put the cans of paint on an old chair, and the painter began to mix the paint. He put a little more red, he put a little more white—it still looked pink to me—and he

mixed some more. Then he mumbled something like, "I used to have a little tube of yellow here to sharpen it up a bit—then this'll be yellow."

What I love about this story is the way it captures both the light and dark sides of the trait. People so curious they must always check out the obvious, just in case, can really hold up a parade. It's obnoxious. Can you imagine how *annoyed* the painter must have been? At the same time, what if he *had* stumbled on something unknown to science? It would have been a big discovery, and Feynman would have made it. And a writer with that same compulsive, magpie curiosity will come back to the office with great material.

I suspect that most people are more curious than they appear; perhaps they are afraid to waste time or look like idiots. If that's you, make a start by noticing how much time you actually do waste—almost all of us do, in fact. We watch television out of sheer inertia. Can't you think of something more interesting than TV to ponder while the coffee brews? I bet you can. Buy a pocket magnifying glass and see what you can find.

If looking like an idiot is your worry, make a start by abandoning your dignity when no one else is around. Read stuff that is over your head or on the fringe. Try new hobbies on for size—learn to bake bread. Go to a storyteller's group. Attend a Go club. In August, watch the shooting stars, and if you like them, take an astronomy course. Take Richard Feynman as your model and let your curiosity out. Go ahead, watch those ants. Look in that dumpster.

In general, it is good practice, whenever you think you know what to expect, to deliberately look for the unexpected, which I guarantee is there, at some level. It will make you a better observer, and you will have more fun. Travel is good—alone, so you meet people. And listen to them. People will *love* to tell you what they do and what they know.

It is especially easy, even for the shy, to go to a big bookstore and sample the magazines, reading a few from worlds you do not know. I mean *read* them, from front to back, including the letters and the small ads. There are magazines for Buddhists, bongo drummers, belly dancers, bakers, chicken farmers, model makers, physicists, snowshoers, curators, CEOs, young CEOs, acupuncturists, housewives, and collectors of kewpie dolls, snuff bottles, or Civil War memorabilia, and it does not matter what you explore. What matters is that you experience many different ways to look at the world, each making sense from its own perspective. Then, when you go talk to people as a writer, you'll find it easier to slip into the other person's view of the world.

Even when a particular intellectual ramble has no direct payoff, you are building dendrites, and you are teaching yourself that nothing, absolutely nothing, is ever precisely the way you expect—which is the scientific attitude in a nutshell. You, too, like the scientist, are learning to stay unsure, with mind wide open to reality.

Curious people make great reporters, since they do not assume they already know. They pay attention, and their eyes and ears and minds are open. It's a fun trait to have; a curious person is never bored. And professionally, the curious can write about many different topics because they suck up so much miscellaneous information.

Always carry a notebook. Pocket or purse, carry something to write on, and carry it all the time. Useful ideas can overtake you anywhere, not to mention useful people, and scratchy notes on the back of a business card don't do the job.

I once sent someone to write a profile of humorist P. J. O'Rourke, a Hopkins alumnus who was at that time editor of the *National Lampoon*. The writer came back with the revelation that, yes, everybody at the *Lampoon* was really funny. The wisecracks flew—and whenever the joke was especially good, everyone in the room would pull out a file card and jot it down. Being funny is a gift. *Staying* funny is a business.

So is remembering what people tell you. Even if you cannot take notes in the moment, it's often useful to write things down the moment you are safely alone, while you still remember the details. So carry a notebook.

Take notes like a reporter. That is, reading your notes should be almost like being there, even for someone else. If you were interviewing me on this very subject, for example, your notes might read as follows:

Hancock interview on taking notes:

- To get effect, take notes almost continuously. Sustained attention tiring, but "u will be so glad u did. U think will remember,—won't."

- Also, impt. patterns of speech, thot, behav. can jump out
  of notes, tho missed at time.
- Key quotes shd. be word for word, in <u>quotation marks,</u>
  so u can tell quotes from summaries. "Key" = v. vivid, v.
  characteristic, or central point.
- "If it's not written down, u don't have it."
- If still in school, take classnotes this way. Enjoy class more
  because awake, grt. skill. "U can tell getting knack when
  yr. frnds. keep borrowing yr. notes."
- On the desk—books, paper, browning banana peel, Yoda
  Pez container. Asked why Yoda—"Is no try. Only do or
  not do."

Create and use a writing space, if only to build your writing habit. You probably know that doctors advise people not to do anything in bedrooms but sleep and make love. In that way your bedroom remains a place with pleasant associations, and you will tend to feel sleepy or sexy (or both) as you walk in, out of sheer habit. Not bad, eh?

Writing works the same way. It helps to have a specific place where you write and only write. After a while, when you go there, the rest of your world will drop away, leaving you free to focus on your words. Do not use your writing space to call up your best friend, play computer games, read the paper, pay the bills, or anything else but write. (Uncle Sam likes it that way, too. To qualify for a tax deduction, a home office must be used exclusively for professional work.)

Annie Dillard says in *The Writing Life* that she favors an extremely plain office, one that elicits no distracting impulse to decorate. That makes sense to me, although I will own to a few sentimental tchotchkes. The room where I write has not been painted within living memory, the furniture is motley, and the floor is bare. But it has what I really care about—my trusty Mac, a big window, a comfy chair, and lots of surface on which to spread out my books and papers.

What do *you* care about? Arrange your space your own way. If you like a setting that is stripped for action, go lean. If you feel inspired by pictures of your family, load up with photos. Use whatever will make you feel good and function well.

That space need not be a "study." In the days when I had to work at the kitchen table, I would transform it into a writing space by setting out my typewriter, scissors, Scotch tape, a cup of pens and pencils, and a pile of nice, clean paper, all just so, in easy reach. Fundamentally, I had a writing ritual instead of a study: I would set the scene, enjoy its order and readiness, then plunge in to mess it up. Even now, when I do have an office, I tidy my desk before I start.

In the same way, you could train yourself to write in your local coffee shop, away from all the distractions of your home.

All that said, do not fuss over your office instead of writing. *Write.* That is so important, I'll say it again:

Above all, WRITE. Writing is what writers do. At parties, people often tell me that they have decided they want to be writers, and they'll get started as soon as they have more time, or when they have their study fixed up, or when they get a new computer, or when they can afford to go back to school, or after the precession of the equinoxes, or something. When I was still an editor, a few would even say they'd get started when they had an assignment, then look at me with bright-eyed expectancy.

Even as party chitchat, these statements seem odd. These same people would never tell a football coach that they had decided to become football players and would begin to train as soon as they got a contract. Would they? I can't think why writing seems different. It's not. If you want to be a writer, write. Keep your day job, but write.

Write about something that excites you, or keep a journal, or find a writer's group, or take a course, or all four together. Write, then get someone to give you a serious critique, then write more and better. If the joy of it outweighs the pain— you're a writer.

Many people, including many who make their living as writers, find it hard to write without some outside galvanizing force—a deadline, complete with someone to whom they have promised a manuscript. That's normal human nature, a need met by courses and writing groups as well as professional assignments. It helps to have a particular set of readers in mind.

To acquire that kind of stimulus, you might scout local, regional, and special interest publications, many of which will take a chance on novices. Write a few sample book reviews or columns about neighborhood news, or gardening, or beekeeping, or whatever you know about. Think up a few feature ideas and go present them to the editor. (Make sure your ideas are congenial with material from the last *several* issues, not just one.) Or try writing a short piece about someone you know and sending it to an appropriate alumni magazine or the local paper.

You may be surprised how helpful some of these unsung editors can be, if you keep an open mind. Not all big league talent is in the big leagues. So look for the small publication that is uniformly thoughtful, peppy, and well-written: Its shaping hand will be someone you want to know.

The reception your samples meet will tell you precisely how good you are *as of now*. That's so important I'll say it again: how good you are *as of now*. A refusal or a piece that gets totally rewritten does not mean you should quit. It means that you tried the wrong market or that you are a beginner. You will soon learn whether you enjoy the work enough to struggle through getting better.

The good news is that no outside authority is needed to vouch for a writer: the quality is there or it is not, apparent upon reading less than a page. So credentials are more a door opener than a requirement—though a writing program does no harm: Working on your writing full-time, with professional feedback, is clearly the quickest way to improve.

If your writing is already excellent in every way, you are as rare as a spotted owl, and some smart person will be happy to collect points for "finding" you. Get out there and hustle.

If you have the good luck to find a mentor (or better yet, to have a mentor find you), seize the chance. Don't insist that the mentor be a perfect human being before you will sit at those feet. Mentors are mostly all too human.

What you get from a mentor that you cannot get from academic courses is a sense of how one capable person actually performs the work, in a day-to-day, already-well-integrated sort of way. If you were working for me, for example, you'd hear most of the things I say in this book. But you'd hear each one with variations, in the context of specific pieces of writing, and in the form of coaching, not general principles. By imitation, you'd also pick up things that I think and do without thought, from long practice, for reasons that are so deeply part of me I can hardly say why, exactly—which appears to be the essential nature of expertise.

Years ago, I heard Marvin Minsky of MIT explain his thoughts on this subject. In trying to construct expert computer programs, Minsky had discovered that experts do not, in fact, follow the rules they will give you if you ask why they do what they do. Rather, experts are persons to whom every case is a special case because they've seen so many that they simply know. Since that lecture, I've several times heard Hopkins wizards of medical diagnosis say something like, "Yes, everything points to X, but I think it's Y"—and be right. Asked how he knew, one such doctor thought a moment and said, "I've heard that song before."

And that's what you get from a mentor. You get to watch the wizard steer by stars he cannot name until after a while you have absorbed ... let's call it an attitude, or a "feel." Certainly it's more flexible than rules. I would say a mentor communicates an approach: a consistent way of Being that in turn gives rise to appropriate Doing.

You can spot the best mentors, like the best parents and the best shrinks, because their former protégés are out there doing the work. They do not hang around being grateful and looking for approval. There is no shrine to tend nor any big anger left from a struggle for independence. The good mentor coaches, then sets you free.

He or she will probably introduce you to a professional network as well, but the network is the least of the gift, because a so-called "network" is really more like a tribe. If you are the right breed of cat for your mentor's tribe, the network quickly becomes your own. If you belong in some other tribe, the network will drift away, or it may never "take" to begin with.

Avoid the false mentor, meaning anyone who insists on "rules" or who is too nice to put you through the pain of growth.

If you lack a mentor, follow good "rules" in a flexible way.
Adopt all usual guidelines, but watch out for the times when they do not quite work. For example, yes, an anecdote is a good way to open. But ... you notice ... this time it feels mawkish. Hmmm ... Through such moments, when you let the material tell you how it wants to be presented, you can evolve a workable state of being on your own.

Keep a journal. Writing programs often require students to keep a journal—a good plan for any aspiring writer. Journaling will help you acquire two crucial habits: (1) the habit of

writing itself, so that it feels natural, and (2) the habit of im-
printing the details of what you see and hear on your mind
long enough to write them down. Memory degrades about
50 percent overnight, so capturing the all-important details
will work best in the evening.

What kind of detail? Well, for example, take the last family
Thanksgiving dinner you attended.

Who was there? Describe them.
Where was the dinner held? Why there? Describe
the place.
Who cooked which dish?
Who arrived first? Why?
Who left first? Why?
Is there anyone in the family who did not come? Why or
why not? According to whom?
Who was always in the kitchen?
Who was never in the kitchen?
Elicit and report two family stories that you had never
heard before.
Report three scenes, with the crucial bits as much word for
word as you can manage, each one to illustrate a different
"truth" about the group. Do not articulate the truth,
however. Let it shine through.
When you describe people, do not describe their clothing,
which our ad-driven culture has conditioned you to do.
When you force yourself to ignore the clothing, you also
force yourself to see the individual.
How did the actual day differ from the day you expected?
How were the two alike?
Smells?
Tastes?
Sensations of touch?
Voices?
Sounds other than voices?
Colors?
Shapes?

You will notice that I am not asking about you and your
emotions, which is deliberate. If you want to write science,
or indeed any nonfiction, form the habit of looking outward
more than inward—though you should know that the inner
you will show up loud and clear, through your choice of de-

tails. It is astonishing but true that a friend or sibling could write about the same Thanksgiving and, apart from names, you might not recognize the scene.

To get closer to science writing, try a public lecture or the county fair or a visit to your veterinarian, capturing the same level of detail but now with less atmospherics, more intellectual content. Keep it interesting.

Don't work hard! That's an order!

Seriously, don't "work." Remember that this journal is only practice, training for your memory and observation. No one sees it but you, and you want to look forward to this time, these final peaceful moments in the evening—well, most evenings—when you call back the day and set it down to remember. Whatever you read, saw, thought, whatever happened, is all fair game.

As the weeks go by, you'll find that you remember details better, or even that sometimes you can play back a sort of mental tape, hearing and seeing the heart of the matter as you write. In its new, condensed form, the event might almost seem to glow on the page. (So *that's* what happened! Wow!)

If you proceed in this spirit of relaxation, you'll enjoy your journal, you'll keep doing it, and you'll grow. If you make it a chore, human nature being what it is ... you won't. It is better to journal for five minutes, if that's all you have, than to skip it and try to do a Big One on the weekend. Keep it regular. If you do have to quit, start again when you can. No strain, no guilt.

Relax. Be serious lightly.

Once you've taught yourself to see and think with some density, you are ready to visualize your reader, focus in on what you want to say, then watch while it flows out your fingers and takes shape on the page. Then you refine it, and that's all there is. You are now a writer, whether or not that's how you make your living.

Finding Stories

TWO

*Luck favors
the prepared
mind only.*
—Louis Pasteur

Now that you have your curiosity unleashed, your eyes
and ears wide open, and your notebook handy, let's try a
little Doing—finding viable story ideas, either for books
or articles, and starting the work.

You'll notice I said "finding," because that's the way to
do it. As discussed in the last chapter, everything is interesting. It follows that everywhere you go, the ground is
littered with excellent story ideas.

Alas, most of us walk right by, often because we have
some preconceived notion about what constitutes a
"proper" story. We think it should be relevant, or have a
human angle, or be certified significant by the *New York
Times*, or perhaps all three. But really, a viable story idea is
much more simple.

**A viable story idea is anything interesting that other
people don't yet know.** You don't need story ideas from
the *New York Times* or any other magisterial source. In fact,
you're better off without. If your idea has been in the
Times, forget it. The world already knows.

Paul Hawken (of Smith & Hawken, the original garden
catalog) writes about this phenomenon in his wonderful
book on how to grow a business. He points out that if
everyone pooh-poohs your business concept—"Why,
whoever would buy garden supplies from a catalog?
People who garden already *have* their tools!"—you may
have a good one. If all your friends just love your idea,
however—"Oh yes, selling fresh cookies in malls and air-
ports, that's really great!"—you've got a loser. The market
is saturated, which is why everyone can see it. They *have*
seen it.

Story ideas work the same way. What you need is something fresh, even unlikely, and you need it live and on the hoof—not in a magazine but in a lab, a new product, a new question, or someone's fertile, teeming mind. Stop reading and go hunting. Talk to scientists. In fact, talk to *everyone*.

Abandon your preconceptions about what a story "should" be. Such preconceptions derive from finished stories, which have a certain polish and shapeliness. But now you're looking for the initial idea, which will be raw, not to say feral. It might be nothing more than a question.

To spot a story, your best clue—*always*—will be a leaping flame in your own mind, that feeling of Oh yes! Wow! Tell me more!

A good idea can come from anywhere. The fertile, teeming mind you seek might be your own or perhaps a neighbor's. (One of my own all-time great story ideas came from the family dentist.)

A young writer recently told me that, on the day after September 11, 2001, his taste in music changed. "I always had the car radio on a rock station," he said. Then the day after the Twin Towers fell, he got into the car and on came the radio, blasting metal rock. "I thought, What *is* this? Bleouch! Now I listen to classical stations. It makes me feel more like I live in an ordered world, with a civilization. You know, a hundred years from now, people will still be listening to Mozart."

I got excited. "Oh! Write about that! That just rings so true, you *can't* be the only one!"

He shrugged and changed the subject, uninterested in his own intriguing self-observation. He was looking for a *big* idea, something with intellectual clout. He wanted something that would make a *book*.

Well . . . maybe. Who knows how that idea would grow, if given a chance? Maybe it's just an Op Ed, or maybe it's a feature, or even, yes, a book. Is Mozart selling especially well? What *is* it about music and the emotions? Music is universal, a part of all human cultures. Hmmm . . .

Whatever interests you—big or small—will interest a reader. Count on it. Then make sure you are open to getting enthused.

First of all, get out of your own way. Remember that you

are not looking for the biggest and best article ever seen on
this planet, an idea that will earn you a Pulitzer before it's
even written. You're just looking for something good enough
to be worth writing and reading—something interesting.
"Interesting" is the sole criterion.

The process is rather like fishing. You put your line where
you think the fish may be, and if something tugs, you tug
back. You don't say, Oh, that's just a nibble, the fish must be
too small. If something tugs, tug back. Find out more. If you
catch enough fish, some of them will be big—and most will
be bigger than you thought at first.

Subjects have a way of opening out, getting bigger and
bigger as you go, not smaller and smaller. Once you understand one subject, it gives you a window on several more,
which you may choose to write instead, or also, or next
week, or years from now. So, in a sense, each story idea is
merely a starting point. If something tugs, tug back.

Clichés can work. Think about it. How does a cliché develop its fine patina?—from overuse, which implies that it
works. Something about it reaches people.

By and large, a cliché is like a proverb: it reflects some ar-
chetypal reality that most people would agree has general
truth and significance. In writing, therefore, you can almost
always make a cliché work, because once you dig into the
particulars of any situation, the clichéd quality evaporates.
Whatever you find will be unique to these people, this situation, this time.

If that's not so, you haven't looked deep enough.

**Do not set out to write a story about the subject closest to
your heart,** meaning material that came to you as a revelation, a bolt of lightning that lit up the entire internal landscape. Possibly this one is the story you've always yearned to
write. It might be the story that brought you into journalism. (After all, it is so important that people understand!) It
might even be the story that everyone except me tells you
to write.

Nevertheless, please hear me out. I had to pay any number
of kill fees (for unusable work) before I finally figured this
one out:

The closer a subject is to your heart, the harder it is to manipulate the material. In effect, you have had a conversion

experience, so that you can only see the topic the way you see it—which is unlikely to be the way a reader (being unconverted) sees it. For that reason, you'll have to struggle even to know what needs explaining, let alone which explanations and examples work. Also, the piece will tend to balloon uncontrollably because you'll want to put in every single precious detail—precious to you, but not to most readers. In short, if your heart is running the show, your judgment will be off.

And that is why I suggest that you put your particular heartfelt subject in the basement of your mind and let it season. Write about it a few years from now, when your perspective will be larger and your skills more developed.

If you cannot bear to wait, you must come front and center with your feelings, perhaps by writing a personal essay. The essay format enforces specificity, and it is so short that intensity enlivens, not discredits—as for Aldo Leopold and Wendell Berry, whose quietly passionate essays have done so much for the environmental movement.

What you must never do is look for a mouthpiece, to let yourself write about your passions inauthentically, under cover of someone else as if you were a neutral reporter. It won't work. Your tone will be inappropriate, causing the reader to smell a rat. In the worst case, you might sound like some unfortunate person ranting in the subway.

For knowledge with a practical application, check out engineers and the many scientists working in governmental agencies and other nonprofit entities. For example, most large cities have a city forester, who will have interesting things to say. Then there's the Veterans Administration, agricultural agencies, social service agencies, and on and on.

For fresh knowledge (i.e., breaking research in both social and physical sciences), look in universities. Corporations also do significant research, but they don't go public with it. Universities still do—even though all the world now knows that basic discoveries can lead to fundamental patents and big bucks. Nevertheless, universities preserve at least the ideal of open scientific discourse.

In this model, science per se seeks to understand our world and how it works—all of it, from the smallest muon to the universe itself. The goal is not technology but pure understanding, in aid of which scientists expect to help one another, in person and by publishing all research. Theoretically, they reveal everything they learn and how they did it, coaching other scholars away from blind alleys. Research thus becomes a worldwide cooperative endeavor, moving fast-forward.

That's the ideal: university science as a shining city on the hill, from which all knowledge flows. Engineers and corporations scoop up the knowledge and apply it, creating technology to make a better world.

Of course, no university is like that (though quite a number in fact surmount a hill). It isn't, it wasn't, it couldn't be. Nothing done by human beings can be so pure and perfect. Yet, until recently, research universities were more that way than outsiders can imagine. The ideal attracted idealists, who did (and do) their best to carry it out.

For example, both the internet and molecular genetics originated in the open era. The original net was not created by entrepreneurs but by academic scientists, with a little funding from the federal Advanced Research Projects Agency. They did it for their own use in sharing data, software, and computer power; the incentive was not economic but intellectual. Likewise, the fundamental genetic discoveries, including the basic technique of snipping DNA with restriction enzymes, were not patented by the people who actually did the work. I remember a reporter asking about patents when Daniel Nathans and Hamilton Smith shared the Nobel Prize for discovering the use of restriction enzymes. Nathans smiled. Patents recognize *inventions*, he replied gently, and no human being had invented genes. Science was a resource for the entire world, and its benefits were not to be restricted by patents. That was in 1983.

The culture of scientific openness has since been modified. For example, certain discoveries get published only *after* the patents are in place. Yet the open tradition prevails, because it does move knowledge forward. In academe, rival labs still share certain materials as a matter of course, and collaboration by e-mail is epidemic. For example, as Comet Shoemaker-Levy 9 splashed into Jupiter, images from the Hubble Space Telescope were e-mailed to astronomers all over the globe, who shared their data in turn. To do it any other way would have been unthinkable. And anyway, is an unshared idea really an idea at all? Maybe not. For scientists, as for the rest of us, ideas spark each other off, showing their full size only as they connect. Sometimes it's hard to know what you think till someone asks the right question or offers one more tidbit of fact.

That being so, every research university remains a happy hunting ground for the budding science writer. There you will find fresh, new knowledge, not only at press conferences or interviews, but also at a continuing stream of short, informal talks. You can find these events by reading the bulletin boards: look for departmental seminars, sometimes described as brown-bag lunches. (It's a lecture, usually in a classroom at noon, Bring Your Own Lunch.) In a medical setting, look also for rounds or grand rounds. If the event is scheduled in an auditorium as opposed to a classroom, a crowd is expected, so either the speaker or the research may be hot stuff indeed. Researchers organize these events to brief one another, there may be four or five on any given day, and they are open to all. Just walk in.

If you want to use the material, however, you must identify yourself to the speaker and get permission. That is not only ethical (because no one knew that the press was there) but also smart: You will want to fact-check, and you will not want to make an enemy. Bear in mind that coverage in even a lay publication may jeopardize the researcher's ability to publish the work in professional journals. You may therefore be asked to wait, which I always did. It's a win-win deal: By holding off, I gained increased access and more time to get my final manuscript just right.

You can also find good material in academic journals or even as near the surface as *Science News*, a slender weekly much beloved by high school science teachers. You can assess the research more clearly, however, in an audience, because you hear the questions and comments of other scientists. When the whole room is excited, you'll know it.

One word of caution: The stories you will find at brown-bag lunches are undeveloped. Normally, it might take five to ten years before they would percolate into full public view, having in the meanwhile accrued workable explanations and examples. At the start, however, both you and your readers may lack the context to really get the new material.

For example, I remember our managing editor at the *Johns Hopkins Magazine*, Mary Ruth Yoe (now editor of the University

of Chicago's magazine), coming back from a lunch to tell us
about a new psychiatric diagnosis. It seemed that there were
people, mostly teenaged girls, who stayed thin by making
themselves sick. And there were people who were skin and
bones yet would not eat because they saw themselves as fat.
Eventually, they could not eat, and they sometimes died. Well!
We sat there with our mouths hanging open. *Really?* They
make themselves *vomit?* But we did not write it up. Who
knew anorexia and bulimia were epidemic? Not me. I also
think we were put off by the taboo.

Along similar lines, back in the early 1970s, a biologist
told me that all the world's important researchers in a particular field had decided to stop work until they could devise
some safeguards. Of course, as you might have guessed, the
research dealt with recombinant DNA, at a time when even
the first primitive methods for snipping DNA were new.

"Re-*what?*" I said. Issues such as souped-up viruses and
Dolly the cloned sheep were before me in that moment a
quarter-century ago. I could have scooped the world, and I
did not even look into it. I was just puzzled. Some scientists
decided to stop their research? Gee. Peculiar. The worry du
jour was radioactivity, and the idea that any nonatomic research might be risky was hard to take in. So I only got to
that story a few years later.

The bigger the story, the easier it is to miss, because the
less it fits any existing mental framework. A truly new idea
demolishes the old framework.

For that reason, once you do find one of these chunky
nuggets, you must take special care to set it in context. Otherwise, even if you can publish it, the article will sink like a
stone, unread. "Re-*what?*" the reader will say, and flip the page.

Anorexia was a hard one, because until the popular singer
Karen Carpenter died, revulsion and incredulous laughter
would have been a normal reaction, for readers as for us in
the office. Yet we could have done it. With the right photographs and interviews, even then, we could have created a
thoughtful, careful, heart-wrenching feature that would have
caused alumni parents all over the country to look at each
other and say, "Could this be our child? She's awfully thin."
And the news would have spread from there. Moving the information out a few years early would have been a public
service, and I'm sorry we missed it.

When you get your chance, perhaps you can do better.

The challenge is twofold: not only to see the important item, buried as it is among other research "news" that will never be heard from again, but also to supply a context that drives the point home. For sure, if you find your mouth hanging open, you have found a story.

You will recognize good science in several ways.

Listen for the cosmic CLICK (as also discussed in the previous chapter with Richard Feynman)—"Well, of *course!* How could we not have seen that, it just makes so much *sense!"* You may feel the click yourself, if you know enough about the field, or the scientist may say it, describing a moment of breakthrough, or other scientists may say it. Whichever, the cosmic click tends to mark a basic discovery, one that will not only hold up but generate future big discoveries, fast and furious.

I observed such a moment in 1973, when Solomon Snyder and Candace Pert announced their discovery of receptor sites that opiate drugs attach to in the brain. Reactions were thunderstruck: "Well of course! That explains addiction! And the body must make such a molecule on its own, or we wouldn't have the brain sites!" One could virtually hear lab doors slamming as researchers rushed off to find receptors of their own.

Track the excitement of scientists, or what you might call the Psst factor, as in "Psst! Are you going to Gary's lecture?" The buzz travels, even the Great Ones come to that lecture, and with luck, you may occasionally be there. If a scientist tells you that "everyone" is excited by something, you should be excited as well, no matter how dull, abstruse, or political it sounds to your untutored mind. Go find a tutor.

Seek the grand simplicity. Here the research has illuminated something that underlies a mountain of complexity, yet is itself very, very simple. Take DNA: Attached along the backbone of that famous double helix, all genetic codes reduce to four, only four, nucleotide bases—adenine, guanine, cytosine, and thymine. The four in turn code for the twenty-one, only twenty-one, amino acids that combine to make up all of life's thousands of proteins. You can see that any revelation about how the four or the twenty-one do their work will tend to illuminate all proteins—exciting, because proteins do the work of life.

Among scientists, the code word for research at this grandly simple level is "fundamental." Think chaos, Avogadro's number, self-organizing systems, and on and on.

Another whole class of story ideas arrives from the other direction: from outside the world of research, in the form of a question, observation, or piece of news. To understand the phenomenon then brings you back to science.

Uncover a detective story (though you probably won't call it that). The puzzle is a sick building, or a sick ten-year-old, or a sick watershed—something, anything, that compels solution. Your heroes puzzle it through, step by step, while the reader tags along. Human beings love to solve puzzles, so much that it must be a survival trait, something that keeps our species going. In any event, if you can embed science or engineering or medicine in a What-done-it, people will read.

For inspiration, read anything you can find by the late Berton Roueche, who exploited this format in the New Yorker as "Annals of Medicine."

Look hard at something ordinary, something so familiar that no one gives it a thought. There's a special delight in learning something unexpected about things we think we know. That forks were not in common use till the fourteenth century, let us say. Or any of the revelations in The Secret House by David Bodanis, who tells more than I ever thought to know about dust mites and the little clouds of dead skin particles that surround us all.

Find special access to the inside story. Everybody loves an inside story. If you can get access to answer the question, How does it really happen? the nature of "it" matters little—and can be scientific, at least in part. Beautiful Swimmers (1976) by William W. Warner is a classic, a book about Chesapeake crabs and crabbing that mingles knowledge from ichthyologists and watermen—especially watermen, with whom Warner must have spent hundreds and hundreds of hours. It shows.

You may well have a special access or special knowledge that you take for granted. Do you? Perhaps you were a toast-of-New-York ballerina till your knees gave out and you turned to writing—but you still know ballet, and you still know many dancers. Perhaps you grew up on a farm, and you can still taste a clod of soil and tell what would grow well on it. Perhaps your favorite uncle is an inventor. Or perhaps he was born deaf and married a deaf woman and had deaf children, and everyone in the family can sign including you ... Well, you get the picture.

Spot what appeals to the visual sense. Today's technology lets us look at everything from the eyes of flies to the birth of stars, from the bottom of the ocean to the everyday flutter of the human heart. Unseen worlds are daily becoming visible.

These stories have special appeal, not only because of their novelty but also because, for most human beings, vision is the primary sense, our dominant way to take in the world; in many other species, vision matters less. Dogs, for example, lack full-color vision but can smell ... Well, I can only begin to imagine. Suffice it to say that a human has 5 million olfactory receptors, while a sheepdog has 220 million. If you were a dog journalist, you'd go around sniffing out the most exotic, nostalgic, rare, and complex smells to share. (Llama doo! Redolent of the pampas!) Since you are human, keep an eye peeled.

Look for something that is showing up as missing, meaning an absence so big it is palpable once you notice. In daily life, consider the experience of driving by a familiar corner and noticing, suddenly, that it is empty: bare earth. "What was there?" you say—and find it surprisingly hard to remember, even though you've driven by five hundred times. But boy, does that corner feel empty.

The memory of what was there comes even harder once the gap has been filled, because our minds adapt to new realities almost instantaneously. And that is precisely the point in time, metaphorically speaking, when a glance back may reveal a disappearance that is well worth recalling. What *was* on the corner? Often the loss was an unanticipated side effect of some other change. For example, Hello computers meant Good-bye slide rules—a trivial shift, or was it? We got a charming essay out of that one, which the author keeps selling to this day: "Elegy for the slide rule."

Here's a more important example: Who has looked at the effects of HMOs on medical *education*, as distinct from treatment? Consider the experience and patient contact that hospital residents no longer have because patients now come and go so quickly and can be admitted only by experienced physicians (the "attendings"). Under the new system, how and what do these new doctors learn from their residency?

Uncover predictions not borne out—a sour sport, like shooting the proverbial fish in a barrel, yet sometimes worth doing. Computers, for example, were initially touted as saving trees. They were going to create "the paperless office," or so everyone said.

I think that the main point of writing these stories is to remind us human beings how poorly we predict, and how much our predictions stem from our wishes. Keep such an item short, or use it to season and undercut whatever prediction is in vogue today.

As a corollary, avoid stories that consist of predicting the future, for reasons you will see if you go find some fifty-year-old magazine predicting how life will be in the twenty-first century. Those stories often display a touching belief in the glories of technology, and they are all wrong, wrong, ludicrously wrong. The one thing we know about the future is that it will surprise us ... though, of course, if thinking about the future draws your attention to something in the now, *that* might be a good story.

As a friend used to say about university politics, "If you must fight an alligator, do it while it's small." Predicting the future is one thing. Pointing out a small alligator with its nose in the trough is another.

Find a special someone. Every once in a while, you will run into someone so brilliant, so appealing, so articulate or charming that you will want to write about that person no matter what. The right personality can make anything readable.

Take Einstein, for example. Do you really think the public much cares about $e = mc^2$? I doubt it, then or now. More likely, readers were charmed by a great scientist who not only fled Hitler's Germany and had a hand in the bomb, but who also let his hair fly free. Best of all, *he would stick out his tongue for the camera*. What could be more disarming?

I am not willing to dumb science down, but I certainly will use any method that helps it slide down easy. It will pay you to take a good, hard look at any charmer you meet in any domain. If not a scientist, is this person a protagonist in some important issue?

The peril of the special someone is that you may become infatuated, a pitfall that is most likely when writing profiles, notably of public figures. I cannot tell you how often I've seen young writers come back from an interview with eyes lit up like candles. "Oh, he's wonderful," one said. Dazzled pause. "He said I asked the best questions he'd heard in years."

All that practiced charisma set loose on a single person can be irresistible—though the reflex cynicism of a more experienced reporter is quite as bad. In either case, the reporter cannot think straight.

Infatuation passes, but if you have to write in the meantime, you may be able to rescue the situation in the following way: Do the best you can, making sure to put in everything your mind winces away from as unbecoming. Then go through systematically and adjust the tone, translating or deleting every single word of overt praise or blame. A "radiant smile," for example, might become "a broad smile" or even "a broad smile that lights up his eyes" (if it's true). Just let the facts and quotes speak for themselves. If taking something out leaves an obvious hole, ask yourself, What did I see or hear that made me think that? Then fill the hole with descriptions, anecdotes, facts, and quotes: no generalizations, just good reporting.

Let the readers be dazzled if they choose—as they often will. After all, this is a special someone.

Elucidate a meaty question or policy issue, which you can address via case study. For example, Can science and engineering find a way to control Nature? When John McPhee asked himself that question, the result was *The Control of Nature*, a book about human attempts to manage volcanoes (Iceland and Hawaii), mud slides (Los Angeles), and the Mississippi. Happily, McPhee spares us the lecture in favor of evidence—some academic, some not, all of it fascinating.

Policy issues normally revolve around a cast of several characters with conflicting goals and needs. For example: the Insured, the Uninsured, the Doctor, the Insurance Company, the Taxpayer, the Hospital, the Residents and Medical Students, the Medical Teachers, the Makers of Medical Paraphernalia ... One compelling story from each vantage point would make a fabulous book, don't you think? Well, for every complex issue there is a fabulous book to be written. Go for it.

The beauty of case studies is that they carry the reader along on the wings of story, to which you can append pods of explanation as needed. At the same time, they anchor issues in the here and now. With case studies, it is not possible to abstract policy from its effects on actual people in actual situations. Best of all, such a piece need not resolve contradictions nor come to any premature Grand Conclusions. You can end on a provocative question.

Questions can be a major contribution to an ongoing societal debate. Once we ask the right set of questions about health care, poverty, drug abuse, and terrorism, we have a chance to evolve some workable answers.

Do not give up just because the story you expected failed to materialize. Maybe there's something even better—a lesson I learned from the late Dick Levine, at one time the primo "color writer" for the *Baltimore Sun*.

Dick surely could find a story. One time, all the local media collected at the Baltimore airport because rumor said the Aga Khan was coming to Baltimore. Photographers and reporters milled about the airport for an afternoon, waiting to greet one A. Kahn, who turned out to be a jeweler from Brooklyn. Mr. Kahn was quite surprised, while the media folk slunk home feeling like fools—all but Dick, who wrote the *Sun's* hilarious front-page account of the debacle.

Absent the Aga Khan, Dick still found a story. He saw the story that was there because he looked past his preconception of what "the" story should be.

Do not confuse a topic with a story idea.

A story idea arrives in a phrase or sentence with a verb, preferably an active one. Something specific happens, in detail. For example: "Why clouds of dirt stop rain falling in the desert." "Agent Orange is still poisoning the Vietnamese, but now it's in the food." (These catchy news heads come from the May 26, 2001, issue of *New Scientist*, always a good read.) Once you can boil down your idea into a specific, active statement, you have the makings of an article. It is this kind of developed idea that you tend to get when you go talk to scientists.

41

*A topic tends to be an abstraction, which you may find by reading
or as your own question about the world we live in. A topic has a large
scale, and it comes in the form of a bare noun*—"fractals," for example, or "prostate cancer," or "AIDS." There is no narrative,
little for the mind to catch on. For example:

Once upon a time, when there was no color printing in
newspapers and very little in small magazines, the occasional
opportunity to print a whole eight pages in color made editors go nutty. Or, at least, it made us go nutty. "Rainbows!"
Mary Ruth and I thought. "Great idea!" So we stuffed our
eight pages with rainbows, both painted and photographed
—rather small ones, so we could fit everything in. George
Washington sits on a rainbow in some Washington lobby, as
I recall. Some of the painted rainbows were scientifically impossible, having the sun nestling into the arch, or the colors
in the wrong order, as we earnestly explained. We also offered instructions for how to photograph rainbows ... what
a flop. We would have done better to find a single focus and
do it right. Instead, we wandered around in the subject, dabbing at it. The pages never jelled.

You can develop books or long articles from a topic by
breaking it down into units, each of which has its own good
story idea. But if you can write only 4,000 words, or 2,000, or
250, you must home in on something specific from the start.

You can spot a mere topic with this question: *Yes, and what
about it?* If the question makes sense, you only have a topic.
Rainbows. "Yes, and what about them?" The answer might
well have become a story idea.

**Every good topic contains a ton of story ideas, which may
or may not be obvious or complete. Most often what's
missing is a specific person around whom to build your
story.** The person is essential because you can never, *ever* create a readable story by rehashing textbooks and articles.
Don't even try.

Topics can be frustrating because you can be so near—and
yet so far. You have dug deep and found something fabulous.
It's interesting, important, relevant, and no one else has
mined it for the public. Yet you cannot get at it. In that case
you must wait, becoming a practitioner of what in medical
circles is called "watchful waiting." In other words, Do nothing now but keep watching, poised to spring into action
when the moment arrives.

Sometimes you are missing the knower, an active re-searcher in the field, whose work must be not only interest-ing but also in the here and now. (Otherwise you'll be re-hashing.) Not only that, the person must be someone to whom you have access.

Sometimes what's missing is the example, as in this story told by Robert Kanigel in his book, *Apprentice to Genius:The Making of a Scientific Dynasty*. The book originated as a story in the *Hopkins Magazine* about what we called a "mentor chain," a lin-eage of distinguished researchers with a particular flair that they pass on to their students, who pass it on to their stu-dents. But what are they passing on, and how? The question made a great story idea, and it also made a great book, and it came about by watchful waiting. Here's Rob telling the story in his introduction:

It was 1981. I had just begun researching an article about a Hopkins neuropharmacologist, Solomon Snyder, who though barely forty had already emerged as an internation-ally renowned researcher. It was our first interview and it was going poorly. I asked him about his discoveries, and he told me. But my questions touched off no sparks. He was courteous and correct, but his replies remained for-mal, his face impassive. I was getting nowhere. And then I remembered. . . .

In preparing for an interview, it is good journalistic practice, and plain common sense, to review what else has been written about your subject in newspapers and maga-zines. Now, sitting across from Snyder in his comfortably furnished office at Johns Hopkins, I recalled something I'd encountered more than once, if fleetingly, in my readings the night before: Snyder, it seemed, had gotten his start as a scientist in the laboratory of Julius Axelrod, a 1970 Nobel laureate in medicine. And so, on a long shot, I steered the interview away from Snyder's scientific accomplishments. "What was it like," I asked, "working for Dr. Axelrod?"

His face exploded with delight. "Oh, it was very excit-ing," he sighed, tone and color in his voice for the first time. "It was wonderful." Whereupon he proceeded to re-count his days with Axelrod two decades before.

The turnabout was startling. It saved the interview. But more important, it left me with a story bigger and more ambitious by far than any I'd envisioned.

Some time before, ... my editor ... had had an idea for
an article about the role of mentoring relationships in science and academia ... and had even started a folder on the
topic, which had just begun to receive scholarly attention.

But that was as far as the idea had gone. The problem
was, how could you get at it, journalistically?

Now, in the wake of my interview with Snyder, I saw a
way to give life to the abstraction that was the mentor relationship. Snyder had been so deeply influenced by this
man, Axelrod, that now, almost twenty years later, the mere
mention of his name stirred him to fond reminiscence.
Here was the perfect flesh-and-blood match to the idea.

As it turned out, no fewer than four generations of Snyder's
particular intellectual lineage were available for interview:
the retired Steven Brody, his student Axelrod, Axelrod's student Snyder, and Snyder's students, Candace Pert among
them. Remember: Luck favors the prepared mind only.

**The bottom line is this: If you find something interesting,
never assume it is not a story. Instead, ask How can I get
at it?—and be willing to wait.** If your mind works visually,
you might think of the process as if you were holding a
big, faceted crystal up to the light. You turn your topic this
way and that till you find a perspective that lights it up—
a story idea.

Finding Out
Research and the Interview

Now that you have nosed out a story idea—or at the least a topic or a juicy question—you are ready to look for the living, breathing person or people around whom to build your story. These are the people you will interview, based on research you will have done in advance.

As a mature working professional, I seldom found scientists reluctant to talk—once they had found me to be well informed and considerate of their time. Writing students tell me, however, that they sometimes get the cold shoulder. Hmm. Well, it helps to have an actual assignment from an actual publication, which students cannot always manage—though that day will come. It also helps to be prepared, polite, and efficient in your interviews, which is the topic of this chapter. But first, let me recommend one more attitude:

The science writer and the scientist are allies, sharing a commitment to science and the public understanding thereof—upon which depend future funding and essential political decisions. Each party brings special expertise to the table. The science writer knows how to translate science for the public, while the scientist knows the science. Whenever you start writing about any particular piece of research, you are entering into a relationship with the scientist, and you will find that it helps—as in other human relationships—if "the deal" is clear. What can you expect from the researcher? What can she expect from you? What do both of you hope to achieve? Each person should have a clear idea of the answers, and the ideas should match. I suggest that you structure the deal explicitly as a collaboration of equals, each having a particular expertise.

This concept will stand you in good stead because it allocates responsibility in a way that makes sense and gives both parties freedom to do what they need to do. The scientist can correct you as needed rather than be "polite" and okay a flawed rendition of the research; you likewise can resist the overly detailed and technical rendition that a few scientists will want to deliver. The scientist need not struggle to write or rewrite for you, and you need not masquerade as a scientist. Instead, you can feel free to ask all the questions the readers will want answered, however elementary, then to translate the result into some appropriate lay version. The scientist need only fact-check.

Note the word *translate*, which my Webster's defines (in part) as "to change the form, expression, or mode of expression of, so as to interpret or make tangible, real, apparent, or the like; to carry over from one medium or sphere (into another); as, to translate a poem into prose, thought into action, or ideal beauty into visible form."

Making science tangible, real, and apparent ... I like that job description, don't you? It is precise enough that we can tell how well we're doing. Defining science writing as "translation" also respects the reader, and it is a concept that scientists accept and understand.

And finally, consider the vexing issue of showing copy. This issue is always live, and more so for students.

Here again, the notion of collaboration helps you out. I usually say, "You will have an opportunity to fact-check, because I want it right just as much as you do. And of course, I will be delighted to hear any other suggestions you may have about the piece." The key word is *fact-check*. Beyond facts, there is no commitment to let scientists rewrite my words under my byline (as distinct from hearing suggestions), or even to literally show them copy. I do make an absolute commitment to get the material right.

On such a basis, showing copy or iffy parts of the copy can work very well. Do it in person, however. Sit right there, saying things like, "We'll say X, then," and leave with the amended copy. If you leave it, the scientist will get second thoughts, and you will be in big trouble. For short, straightforward stuff, read the iffy bits over the phone.

Before any interview, do your homework. Do not go to interviews unprepared. In fact, do not so much as make a phone call in a state of total ignorance, lest you get found out. As in other relationships, first impressions matter.

The scientist has no responsibility to make the material

simple enough for you. Some are good at it and like to teach, but if so that's a bonus, because it's not their job. It is your job to master whatever you need to master. Sometimes mastery will be a piece of cake, sometimes not. It can be done, however—and I say so as a former English major, who nonetheless managed to write about everything from molecular genetics and chronic pain to the birth of stars. The secret is to start with the Gestalt—the big ideas that structure a discipline, so that you have a mental framework on which to hang the details.

Neophyte writers often stumble because they think preparation requires knowing it all. For example, they might settle down with an encyclopedia and try to bone up on *all* the functions and interactions of *all* the immune cells, including those not pertaining to the particular research, all in one afternoon. That way lies confusion, not to say despair.

Instead, start by making sure you grasp the basic level, and I do mean basic: ("Antibodies tag material for other immune cells to attack.") Like that: *basic*—material that might appear in a good, family-type encyclopedia. Take notes for a cheat sheet if you need one; writing is an open-book test.

Then when a scientist talks about such-and-so antibody, at least you know what antibodies in general do, so you're halfway to understanding why this particular antibody matters. You can follow the train of thought. Later, you can pore over textbooks or the journal article the scientists wrote for their peers, looking up any unfamiliar word you run into more than once, and *really* get it.

At every level, form the habit of asking yourself: What is the central idea here? Such a focus will help you learn, and it will help you write.

As a rough guide for what basics to home in on, go online and dig up the abstracts for previous research by this particular research group. Can you follow the abstracts? Do you have a rough understanding (with cheat sheet in hand) of each article's key words? Look things up until you can and do. Or use good common sense. For example, when pieces of Comet Shoemaker-Levy 9 were bombarding Jupiter, the Hubble Space Telescope had the world's best view, and its headquarters are on the Johns Hopkins campus. But before I went over, I spent an hour boning up on the solar system, especially comets and Jupiter. It was enough.

You are ready to start interviewing at the point when you

know enough that you won't waste the scientist's time, yet the questions a reader might have still tickle your curiosity, too. You will stay on point and elicit good, lively quotes that way.

If you have a current press release, so much the better. Read it with extreme care. Press releases vary from superb to awful, but even the worst have one definite asset: You can be sure the scientist approved it, probably after a careful reading. Therefore, you can use the press release to answer some of your own questions. Can you write "prove," for example? If the press release uses the word, you can—and prove is a big word, to scientists. If not, not, and you may want to ask why. What more would be required to constitute proof?

Are there places where the language suddenly becomes finicky, dancing delicately along a knife edge? Hmm. When something is being written around, it will pay you to wonder what and why.

An important note in passing: Whenever you take notes from written sources, including press releases and encyclopedias, take the time to paraphrase as you go. For one thing, it's a good way to test your comprehension; if you cannot rephrase the idea, you didn't get it. Second, you don't want to lose your Pulitzer because someone discovers you were plagiarizing. "I downloaded it from the web as background and forgot it wasn't mine" would be a lame excuse.

If you think you might want the actual words, keep them *as* a quotation, using quotation marks and restating the source (because pieces of paper do get separated). Then double-check to make sure you have it right.

Whom to interview: As a student, start with what and whom you find, rather than aiming to interview a Big Name on the latest topic to have adorned the *New York Times* science section. For one thing, you'll meet with fewer No's. For another, you'll have something fresh, even though it may be small. This advice applies to all writers, not only novices, because in my experience the best stories are always found, not manufactured.

Stay awake! Quite apart from brown-bag lunches and press conferences, stories crop up everywhere. As an active professional, sometimes I'd find four or five in a day. Other days,

I'd find none—and the difference was in me, not the day.
Even now I see stories everywhere. For example, yesterday I
went on a house-and-garden tour in downtown Baltimore,
including a site where a group of young archaeologists were
digging up nineteenth-century latrines behind what had
been Baltimore's earliest incorporated synagogue.

Don't you think there's a story in that dig? I do. There's a
story almost everywhere, and every small story can open out
into a bigger one. Let serendipity happen.

**As a student seeking interviews, you should know that
people find it much harder to say No in person, especially
if the request is a modest one.** If you approach the speaker
after a brown-bag lunch, she will know you're not fishing
blindly and that you already know something about the subject ...And after all, you have just sat through her lecture,
looking bright-eyed, and you are not asking for much—an
hour over sandwiches, perhaps, you to bring the sandwiches.
Everyone has to eat lunch, right?

I have also interviewed people on their car phones. We'd
have an appointment for their drive home one evening; the
scientist would call me once he got safely by the worst of the
traffic. Or you might ask for time "with you or any of your
associates," especially if you are only fishing. You don't need
the high honcho for a basic briefing; an eager young associ-
ate may do even better.

Interviewing is an art, and one you will mostly learn by
doing. But I can promise that, if you are well prepared and enthusiastic, even your first few interviews, however lurchy, will
give you the material you need to write. How can the scientist resist? *You genuinely want to know something very close to his heart.*

Plan to do your interviews in person, at least initially. Scientists communicate with each other by e-mail, and they may
suggest you interview that way. It sounds convenient, right?

Wrong. You might try it as a last resort, perhaps if the
other person lives in India, but in general it's a poor idea. An
e-mail so-called "interview" will necessarily be herky-jerky,
the product of many separate days and moods and contexts,
at both ends of the line. If you don't sit down together, how
can you develop any authentic train of thought? How can
you generate trust and connection, the indefinable juiciness
that lets people work together well? How can you write

without the visual details with which we human beings orient ourselves?

E-mail can work well for follow-up, or if all you need is a brief expert reaction to someone else's research, but the e-mail "interview" tends to devolve into mere information: dry little packets of fact. I find that a piece written that way acquires the hollow, depersonalized sound of an encyclopedia entry, and I would abandon a story idea rather than depend on e-mail interviews.

Telephone interviews may be a feasible alternative. Personality transmits on the phone, and you and the scientist can develop significant rapport and a coherent line of thought.

Interview in the morning, on the other person's turf, and never in a restaurant. Timing is no deal-breaker, but if you can, avoid the midafternoon droop: make your appointment for the morning. You'll get a better interview when both of you are bright-eyed and bushy-tailed, as my mother used to say, and when the day's distracting little crises have not yet arisen.

Never interview in a restaurant. Silverware clicking on plates and the conversation at the other tables will obliterate every sound on the tape, nor are you well situated to take notes. Distractions abound.

At the person's lab or office, by contrast, things are quiet, the scientist feels comfortable, and serendipity can happen. If the subject widens, you and the scientist can scamper down the hall to look in the lab or talk to someone else. If there's a reprint or photograph the person wants to give you, you're in the right place. (What you go home with that day, you can be sure you have.) Because you two can see each other's faces, you will avoid many small misunderstandings, and both of you are more likely to venture a joke or a wild idea that suddenly comes to mind. The conversation can flow in a natural, easy way, so that the resulting piece will have a certain indefinable flavor.

Even when you interview someone several times, the best quotes often come from the first session. For that reason, you should be especially conscientious in preparing and conducting a first interview. Tape, take notes, and generally do it right, because the truth is, people cannot help but peacock at a first meeting. They'll be adrenalized, so they will ruffle out their feathers and speak with a little extra punch.

As for you, you need to capture that peak while it's there because, as social animals, we are trained not to repeat our stories. If you have to come back and say, "Tell me that story again about the time ... " you will elicit only a pale ghost of the original.

Leave yourself some extra time. If the appointment you were given is from three to four o'clock, the scientist may well have another meeting at four—but you mustn't have one. There's always the chance that she'll want to keep talking or to show you something in the lab. If so, you'll want to stay.

When you make the appointment, do describe your purpose and "the deal" (if asked), but do not send questions ahead, for two reasons: (1) because you want fresh, spontaneous answers, and (2) because you do not want to limit the interview, a priori, to only the specifics that you knew to ask. You want to leave room for the new and exciting. "I'm sure you know anything I'll ask right off the top of your head" can be a good way to say it, because it is so patently true.

A brief description of your purpose should be enough. For example:

"To talk about your work on protein folding for a news item in the *Weekly Blather.*"
"To talk about the implications of your work on protein folding, possibly for a feature article in the *New York Times.*"
"I understand you are a gatekeeper, one of those people who always knows what's going on. I'd like an hour at your convenience to hear about whatever is exciting people in the field right now."

Even though you did not send questions, you will sometimes find, when you get to the appointment, that the person has prepared a speech anyway. In that case you should listen. Sometimes people must fulfill their own agenda before they can pay attention to yours.

If their agenda is incompatible with yours, you will need to be gentle but forthright: "Not of interest to our readers" is the time-honored way to phrase it. Or you can blame the editor: "I will discuss your idea with my editor [my teacher], but I know that what intrigued her was the protein folding. Perhaps we could go on to that now?"

And—sometimes there's no extraneous agenda. Sometimes the scientist knows better than you what you should be asking. He's just jumping the gun, rushing ahead into exactly that new, exciting stuff you were hoping to find. Don't cut anyone off prematurely. *Listen thoughtfully.*

Important: Make sure, sure, supersure that you do not mislead the scientist into expecting more than you have to give. While most people like to help, they like it more if they know that is what they are choosing to do. So if you are scouting for stories or background rather than definitely writing about this particular person's work, make it clear. If you are writing a news item as opposed to a feature, spell it out. If you are a student working on a paper, do not masquerade as a full-fledged writer with an assignment.

Publicity is not a high-priority goal for scientists, especially those in academe or governmental agencies. As a group, they care only about the opinions of a few illustrious persons of whom you and I may never have heard. They may begrudge time that gets them a lot of publicity, yet not begrudge time in which they are basically teaching you, investing time in creating a knowledgeable writer.

If you get such a gift, be courteous and grateful. Keep the door open, not only for yourself but also for future students. And remember that a thank you note on e-mail has less impact than an actual handwritten note (though e-mail is better than nothing).

Prepare your questions ahead of time and write them down. That is not to say that you will *ask* them as written. If the interview goes well, after the first few questions you'll be having a lively conversation, and you won't even be looking at your questions. You'll be making eye contact, with an occasional glance at your rapidly scribbling hand, and what comes out of your mouth will be a direct response to something the scientist said. You'll be tackling the subject in an order governed by his train of thought and in language that reflects his—in short, your questions and comments will be better than what you wrote down.

Nothing good will happen, however, if you do not prepare coherent questions and write them down in some sensible order. Once again, the wisdom of Louis Pasteur applies: "Luck favors the prepared mind only."

In preparing your questions, stay simple and straightforward, like Bill Moyers. Your purpose is to elicit the lively explanations you need for the piece, not to impress the scientist or to fill in gaps in your own education. Doing your work well is the best way to be impressive.

Many inexperienced writers are afraid the scientist will think they are dolts, so they work up long, elaborate questions, the sort of scenario-setting stuff that looks well informed on paper. Don't do it. The questions you see in printed interviews were usually written for structural reasons, to make the interview come out sensibly after the transcript was cut and rearranged. No one actually said anything so long-winded, as you'll know if you try to speak one of those three-thought mindbenders.

If you are afraid the scientist will think you are a dolt, you can always say, "Our readers will want to know ... etc."

For starters, make sure your questions cover the newsroom's famous five Ws—who, what, where, why, when. Then add another W for Wherewithal (the funding, as discussed in chap. 1), plus an H for How and a big question mark for Why Should the Reader Care Anyway? WWWWWWH? Even when you know the answers (and I hope you do), you want much of this basic material in the scientist's own words, which will always read better than you regurgitating from an encyclopedia.

You may wonder why you should prepare if you're going to ask the basics anyway. Well, because preparation not only reassures the scientist, it also lets you get over the ground at speed. If you're prepared, you'll know when you have enough on any given aspect. (Okay, that's a great quote, I can move along.) In that way, you'll have lots of time to dig into whatever seems fresh and full of panache.

So. The five Ws (plus W plus H plus the big question mark):

Who usually will be a team, occasionally from multiple universities in multiple countries, and you cannot list all those people. It is a pleasant courtesy, however, to credit all principal players, and you must find out who they are. Normally, that will include one or two graduate students or postdoctoral fellows who actually did the bulk of the hands-on work. If you have time, talk to these young folks, too. The researcher will not mind. In fact, the better the scientific team, the more the leaders seem to want to credit the junior members—who are their future colleagues, after all. They are also your future sources, come the day when both you and they are distinguished in your own right, and they have much to say already. Why not meet them now?

With regard to *What*, make sure you know where the background explanation stops and this particular research begins. You'd be surprised how easily that line can blur in your mind.

Why and *When?* Why this line of research and not some other? Why now and not before? Guaranteed, there must have been other ways to approach the issue, so what was the advantage of this one? The answers are always part of a larger picture, about either the science or the research strategy or both.

How, precisely, was the research performed? You want to know at about this level: "We do this because A, then that to create B. Then we put the B in the glomerator and wait. What we hope will have happened after 24 hours is ... " Or, "We assembled a control group of 230 people. The group had to be that big because X. It was important that they all Y because Z." The root of the matter is implicit in these mundane details, so knowing them will clarify your thoughts. Occasionally, you will even want to hang the entire piece on a narrative structure: "The team wondered X, so they did Y. Unexpectedly, results were M. So they started again, doing H." And so on.

"Why should the reader care?" can sound hostile and you won't want to phrase it that way. I usually say, "What are the long-term implications of this work?" A friend of mine likes to ask, "If you had to state the significance of this work in twenty-five to thirty words, what would you say?" He and the scientist sit there and work on the statement till they get it just right, a process they both seem to find fun and illuminating.

Imagine yourself as the reader. Seriously: take five or ten minutes and work into the role. Then ask yourself what you as reader already know (or think you know), as well as what you need or want to know. Questions generated this way will be qualitatively different from those of a walking head. They will elicit far better answers—especially if the subject is medical. For example:

<u>Detached, the questions of a walking head:</u>
What are the initial symptoms?
What is the intermediate phase?

What is the usual treatment?
Does the medication have any adverse effects?

<u>Questions from an imaginary patient, parent of patient,
or friend of patient:</u>
What should I look for?
How can I tell if we're getting good care?
Where can I get the best information?
What are the trade-offs on all the treatments? She's always loved her long hair. Is losing her hair worth it?
What can I do that will help?
How will I know it's time to call 911?
When doctors and nurses get this disease, what treatments do they choose?

Give particular thought to your first question, which has several jobs: it should start the conversation off in the right direction at the right level, *and* it should be a big fat juicy one, right over the middle of the plate—something the scientist can hit out of the ballpark. You want her to feel satisfied with both herself and you. ("Oh yes, she's okay *and* I can handle this. I'm hot today!")

The opening question should not be personal. While Barbara Walters asks people things like "What kind of tree would you be if you were a tree?" the rest of us find it works better to go straight to the science.

Simple can be best, especially if time is short. For example: "I have studied the material you sent me, and it seems to me the heart of the matter is [FILL IN THE BLANK]. Is that correct? . . . How would you phrase it? . . . How did you get the idea?" And you're off. Don't try that one unless you are dead sure you understood the material, however. Your purpose is to simultaneously reassure, indicate the level of discourse, and cut to the chase, not to expose yourself as unprepared.

Curiously, it's okay to be semi-informed so long as you show you're aware of it. "I have studied the material you sent me and I got stuck on X, as I think the readers might. I don't understand [WHATEVER YOU DON'T UNDERSTAND, IN A CRISP TWO SENTENCES]." This formulation also cuts to the chase and indicates the level of discourse. It's efficient. It is even somewhat reassuring, in that you show yourself to be a person who thinks with enough precision to know what you do not know. In such a case, you will often turn out to be missing some large, basic chunk of knowledge; the chunk did not show up in key words because it is taught at the level of Physics 101 or Basic Anatomy, then taken for granted. Aren't you glad you asked? Now you have that missing background to give the readers (or to artfully write around). The next time you cover this discipline, you'll be that much better off.

An omnium-gatherum of question ideas: The scientist's curriculum vitae or resume may well rouse your curiosity. For example, you see from the list of publications that this person has been pursuing the same line of research for ten years. Must be she thinks it is really important! Why? Or if she has recently switched, why? What is compelling to her about this new question? Or is there an interesting connection between the old and the new work?

If time allows, you may want to drop your line in deeper waters: If you were starting your research career again, knowing what you know now, which area would you target? How did you happen to enter your present area of research? What are people in your field excited about right now? What do you think is the most exciting issue in all of science today? Of all the scientists you have known, which one do you admire most and why? As I write this paragraph, those questions as a group ring hollow. However, such a one by itself, when triggered in your mind by something the person said, can work well.

And finally, here are some generic questions that you may be able to adapt.

> Did you have a Eureka! moment in doing this research?
> What new possibilities do you now see?
> Do others in the field generally accept your data? Do they generally agree on your interpretation? What is the central issue?
> What was the biggest surprise you got in this work?
> How did you first get the idea?
> What comes next? Is the next experiment already clear to you?
> What is the next important question?
> Do you have any intuition about what the answer might be?
> Ultimately, where do you expect this line of research to go?

"Do you have any preliminary results?" Once you've launched into implications and future work, look for something concrete. Scientists almost always know more than they can prove, because to get the next grant, they must offer evidence that what they propose is worth doing. In effect, then, much research funded by grants actually nails down results from preliminary work that the team had already done, late at night or sketchily, on bootlegged funds. A full experiment will not necessarily confirm preliminary results. Even if it does, the most prestigious academic journals often refuse to print what has appeared elsewhere—yes, even in a local newspaper. From the scientist's point of view, then, prepublication blabbing carries a heavy price.

Nevertheless, you may be able to elicit a hint. Scientists love to talk about their work, and once one has described the wonderful question she's seen, she will want to say more. So, if you develop a reputation as a person who can keep your mouth shut, you may hear a lot more—material you cannot yet report because it is too preliminary. But you can use it as a basis for informed speculation. (If this-and-so is true, that could imply this-and-exciting-that.) Also, you can use early information as a basis to ask, "When will you be able to talk about this idea on the record?" You may have a pending scoop.

In writing about a disease, ask doctors how patients commonly describe each symptom. Take depression: "I don't enjoy things anymore" and "It's like the whole world is gray." Both phrases are more evocative than "flat affect," don't you think? Or compare "a feeling of pressure on the chest" with "It feels like an elephant is sitting on my chest." You can use patient descriptions to amplify the diagnostic labels: "a sensation of pressure, which many patients describe as feeling 'like an elephant is sitting on my chest.'"

Along the same lines, ask doctors, "What do you tell your patients about this possibility?" In general, whatever she says to patients will be a good quote: well honed, clear, and direct—of course, because she has been practicing that explanation for years.

To ask often and everywhere: "Can you give me an example?" Or, "Can you give me an analogy?" Or, "What do you think is the best example of that?" The human mind seems to store things in sets and subsets, the efficient branching structure apparent in many a course syllabus. Sets may be quite abstract.

We do not learn by abstract sets, however. We learn by starting from something concrete that we know or can imagine, to which we attach the new idea. Your reader is learning from your article, so you must start with the familiar. You might start explaining surface tension, for example, in terms of the familiar water spider, which appears to skate upon the surface. How can it do that? Well . . . thereon hangs the story.

The scientist you are interviewing is long past needing any such crutch, of course. He can visualize the abstractions with no problem, so may not think to give an example or analogy. In that case, you must ask for them as needed.

If an analogy occurs to you as you listen, you might offer it and ask if it works. In general, however, the best analogies will come from the scientist because they derive from a grandly simple view of the field. The scientist may have such a view, but you almost certainly do not.

"Compared to what?" Hold this question at the back of your mind, so it will automatically pop out as required. Let's say such-and-so operation produces a 70 percent survival rate after five years. Great! How does that compare with other treatment strategies? With similar operations for other cancers? Is the difference statistically significant?

"Tell me more" is a magical, all-purpose question because it is open-ended, giving the speaker permission to take the idea wherever he or she wants. It's a good way to go deeper when you can tell something is important but don't yet know why or how. You can also use this phrase to redirect the conversation, if you need to, in a way that will not seem jerky or rude: "I'd like to come back to this, but first I'm still wondering about something you said earlier [INSERT WHAT-EVER IT WAS]. That sounded important (intriguing/unexpected/whatever) . . . Tell me more."

One last question: "Who else should I be talking to?" One fertile, knowledgeable, enthusiastic source will be your best lead to another, because the good guys know one another. But stay courteous. Be sure to get permission, in some form that mirrors what you would actually say: "Is it okay to say you suggested I call?" "Is it okay to say you thought she might like to talk to me?"

When you have such a referral, even a Big Name may give you time, even when you are still a student.

Bear in mind that referring you on is taking a chance, albeit a mild one. The scientist does not want a reputation for siccing dolts on other people—but gains points by spotting a comer, someone the scientists will later be glad they know. If you get treated like a comer, take it seriously. Follow through and make sure you justify your sponsor's risk.

On the day of the interview, be prepared in all the common-sense ways. Make sure you're on time; that your tape recorder works; that you have enough pens, paper, tapes, and batteries; that you look okay. And—these next two are so basic they're easy to forget—make sure you have eaten and that your bladder is empty. Few things are so distracting as hunger pains, which is why you might keep a nutrition bar handy. It can save your bacon.

Dress appropriately, meaning professionally but not what used to be called "power dressing." Your goal is to make the other person feel relaxed and comfortable, not intimidated or even much impressed. Dazzle 'em with your mind and your writing, not your fashion sense, especially if you are a woman. Keep the transaction professional.

In an academic or computer-lab setting, professional attire may be very informal indeed—for the natives. As a visitor, your style of informality needs to stay respectful. For men, a tie and sports jacket (even if both are later taken off) will not hurt. For women, the specifics of "appropriate" vary wildly over time: do the right thing.

Having taken care of all important trivia, you can now begin the interview with no need to think about yourself. Rather, your undivided attention will be available for the other person and the fascinating things you are about to hear. As you walk through the door, mentally give your troubles a kiss and skootch them over. They will wait.

Once in the room, there's no rush to start. Allow a few moments for the two of you to get used to being in the same room, the process that I call "dog-sniffing." Dogs need to sniff each other, and people need to ask whether you had trouble finding the building, is it still snowing, where should we put the tape recorder, do you want any coffee? … It's all dog-sniffing, two to three minutes of adjustment time. (I have sometimes wondered whether in literal fact the need is to sniff each other, but that's another story.)

Whatever happens while you're chatting will be fine so long as it is cordial. However, do not get so relaxed that you lose control of the set-up and find yourself talking across a desk or from too far apart (too far meaning more than six to eight feet). You need to be close enough that intimate conversation is possible. To a scientist, talking about her work may indeed be intimate.

You'll know that it's time to start by a tiny, expectant silence. Then—GO. Deliver that big fat pitch.

Follow the thread. As a culmination for all your carefully prepared questions, you should now … ignore them. Well, not quite, but close to it. Once you've launched a topic that intrigues both you and the interviewee, be prepared to follow the other person's lead. Metaphorically, you are dancing, and if your partner dips, you should dip, too. Your questions should now arise naturally out of what was just said, in a process you can think of as following the thread.

The thread: Imagine you're in a wood, this person's mind, and the topic you have opened up constitutes a thread. Pick up the thread and follow it. Every once in a while the thread will lead into a clearing, where you can see several other threads coming in to join. You may want to pick one up, or you may want to stay with your original thread, or you may want to roam around the clearing. If in doubt, choose whatever seems to excite the scientist. Go for the juice.

Letting the interview flow can be scary at first. It can feel like "Oh my God, I've lost control." If you suffer that fear, look to see whether you're getting good stuff. If you are, you're doing fine. If you're not, you can always return to your prepared questions.

LISTEN. Do not be afraid of silence. It often means that the interview is going very well indeed: you've got the person thinking, not answering by rote. If she's talking along, then trails off and gazes into space … keep your mouth shut. Let the wheels whir. The next thing out of her mouth is likely to be a nugget of gold.

Remember, too, that a person who is thinking may have no idea that the silence has stretched for a minute or more. If you come in with a new question while she is still processing the old one, she may feel you are interrupting. If you "interrupt" her twice, you've blown the interview.

If it appears she's gone off so far she no longer knows you are there, it's still best not to break in with a new question. Speak quietly. Say something like, "You look excited [worried/thoughtful]. What are you thinking?"

Along the same lines, be slow to assume your question was misunderstood or that she is evading you. There is a certain type of mind, not uncommon in scientific circles, that always gears up with a pause for thought, then lays out the data, and only then states a conclusion. The sequence goes about like this: "[Pause] ...X is true [pause] ...Y is true [pause] ...Yet at times M is also true [pause] ... Overall, I would say that ... " and here comes the answer, five to ten minutes after you asked the question. And what an answer it can be, rich and subtle. I sometimes feel as if the scientist has assembled me a bouquet. Pick a flower, pick a flower, pick a flower, add something unexpected, and suddenly ... an object of beauty.

In print, since most readers will need the anchor of a general statement, you may want to move the grand finale to the front. In the interview, however, you should not press for it. If you wait, it will come. You'll know the bouquet is complete when she stops and looks at you with the expression of one who has just uttered the punchline.

At that point, if you did not understand, you must say so. "I don't understand how that relates to our topic. Am I missing something?" (In passing, notice how different "I don't understand" sounds from "You didn't say.") She may have assumed the conclusion could go unspoken, as obvious to you as it might be to a colleague.

Among researchers, "I don't understand" is an honorable admission, one that is made daily, because it is the basis of all scientific investigation. To ask the right question, you must first realize that you don't understand.

Be alert for body language, both your own and the interviewee's. There are any number of good books and videos on this subject, but the basics of reading people we all knew as infants. I'll just remind you:

Do not cross your arms over your chest, and be alert if the interviewee does it: it is a transcultural signal of rejection. Face the person, your body open in a signal of reception.

Never lean away. If anything, tilt forward.

The conventional advice to meet the other person's eyes is good but can be overdone. Do not stare like Dr. Mesmer or

Charlie Manson, and do not be put off if the interviewee
sometimes appears to be watching an internal blackboard.
Once you've got him thinking, an internal blackboard may
be exactly what he needs to see. The red flag is a pair of eyes
that *roam*, yet never meet yours.

If you feel uneasy yourself, or if the other person seems uneasy, check it out. "You seem a little distracted. Is something
wrong?" "Distracted" is a good, emotionally neutral word, and
often accurate. One woman said, "Actually, I'd like to make sure
my fifth-grader got home. I usually call home about now."

Well, that was easy to solve.

If the problem stems from something you said or did, the
sooner you catch it, the sooner you get back on track.

Use a tape recorder, but take copious notes. Nowadays, all
journalists tape. But in the old days, only TV and radio
people did. We writers just took notes, and we got *good* at
it—because we had to. We had no other backup. Which leads
me to a confession: I do tape. Of course. But I seldom make a
transcript, and when I do it's partial. The truth is that my
notes are better to write from—more clear, more lively,
more thoughtful, and the lord knows more succinct.

I have several times had the opportunity to compare notes
with transcript, both my own and those of other people, and
I conclude that, in taking notes, we filter out garble as we
go—*lots* of garble. (Very few people speak in coherent sentences.) At the same time, much nuance from facial expression and body language seems to leak into the words as they
move through the note-taker's mind and out the pen. I do
not think of myself as editing the quotes when I interview. I
experience myself as writing down what I hear as exactly as I
can. Yet the quotation as it appears in my notebook is seldom
precisely what the man said. It is more like what he *meant*, in
his own characteristic cadence but cleaned up and somehow
more clear and forcible. No one ever accuses me of misquot-
ing, and other old-timers report the same experience.

This effect arises after years on the job, and you will
doubtless want to tape. Do take detailed notes, however, if
only in case the machine fails. Get good at it. If you can rely
on your notes for everything but the most complex and
technical details, you will have juicier quotes and you will
save yourself hundreds of tedious hours transcribing.

Also, taking notes in detail will force you to listen actively, so that you will seldom go home and realize you missed several big openings (a nasty experience). Any fool will notice when the bird of thought takes wing—but if it only hesitates on a brink? That's harder to spot. If you are listening closely enough to take detailed notes, however, you have a good chance to see the opening and precipitate the take-off, either with something you say or with your inviting silence.

You will get the best of both worlds by listening to the tape one time only, to flesh out your notes—though you may be surprised at how little you must add or change.

Taking notes on your laptop will be an extreme temptation if you type fast, but I recommend it only for follow-up phone interviews. When I tried it in face-to-face interviews, the machine took enough attention that my rapport with the other person suffered, plus I often lost track of the content. The problem is that, when I'm typing at the speed of speech, the words blur into a succession of letter combinations that flow straight from ears to fingers, with no detour to the cerebral cortex. So, to know what I've just typed, I have to read it, like anyone else.

I can do that if I'm on the phone. If I also have to make eye contact and maintain rapport (let alone think), I cannot. For that reason, I believe that hand-written notes still have the edge.

Do not, repeat not, sort "significant" from "insignificant" as you take your notes. If you do, either fatigue or your preconceptions will do the sorting, and they often will be wrong. Aim to get it all and sort later.

Scribble more or less continuously. In that way, the interviewee will soon forget it: Your activity becomes one more ignorable item, like the drone of the air-handling system. If you stop and start, however, the other person cannot help but notice—and may become self-conscious.

Do not absent-mindedly fall into your college habit of writing down only material you do not know. The way this particular mind fits the pieces together is what you need to record and convey.

Distinguish actual quotations from your own summaries by careful use of quotation marks. Later, when you write, you will weave the quotations into a tapestry unified by your own clear, unobtrusive prose. Weaving will avoid monotony and take you over the ground fast, yet give the reader an experience of the speaker(s). For that purpose, what you need is a range of quotations that are extremely clear, vivid, humorous, characteristic, and/or precisely on point. When a good one flies by, get it as exactly as you can, and mark the exact portion with quotes. A three- to four-word unit that is just right is as worthy as an entire sentence or two. Are there particular nouns and verbs that this person uses again and again? Jot those down, too.

If something amazes you, rouses emotion, or makes the penny drop so that you suddenly understand, mark it in the border with a star. It will probably affect the reader in the same useful way.

Note down mannerisms, interesting objects in the room, and any kind of action, because all have meaning. We may not know the meaning, but if the speaker takes off his glasses, starts to pace, gets up and looks out the window, etc., etc., it may be worth reporting; it will add dimension. If the action is not in your notes, however, you won't remember the timing.

Leave some blank space on your pages, perhaps a big left margin. You may want to write in questions and comments later, not to mention a few tidbits from the tape.

It will help you write notes at speed if you memorize and use a judicious assortment of standard scientific and mathematical abbreviations. Such as:

$\therefore$	therefore
$>$	is greater than
$<$	is less than
$\sim$	is approximately
Na	sodium
U	uranium
$\rightarrow$	becomes, leads to, etc.
$\neq$	does not equal, is different from, must be carefully distinguished from, etc.

Double-check your understanding as you go along. Say things like, "I think I hear you saying that ... " Or, "Would it be correct to say that ... " Or, "That sounds to me like [analogy]. Does that image work for you?"

If you keep checking, both you and the scientist can be assured that you're on track. There's no shame in not knowing—your expertise lies in how to translate research for the public, not in the research itself. Furthermore, what confuses you is likely to confuse the reader, so any correction will be useful. And if you've hit on some useful analogy, the scientist may be inspired to take it further.

The best correction I ever elicited was from an economist named Carl Christ who said, "Oh, I see! You think economics has to do with *money!*" Now there was a thread worth following.

If something the scientist tells you does not jibe with the background material in your head, wake up. Probably what you thought you "knew" is old information, now known to be false. Say something like, "Hang on a second. I have always understood that such-and-so. But what you say implies this-and-that. Tell me more."

Make interested sotto voce noises of any sort congenial to your personality. People need to know you are with them.

I tend to murmur things like "Wow!" or "Intriguing!" because I'm an enthusiast, but something else might work better for you.

There's only one taboo: Do not say "uh-huh." Some people take that as meaning, "Oh, I already knew that." If so, it will stop them cold. The only way to not say "uh-huh" in an interview is to not say it in ordinary life. If you have that habit, break it.

When the interview is over, even though you are prepared to stay, don't be hard to get out of the office. About five or ten minutes before the appointment is scheduled to end, say something like, "Time's running out. Let me look at my questions to see if there's anything we've missed." When you look, you'll probably find you've covered it all, albeit in a different order, except a few questions that now, in the light of the interview, were clearly off the mark. And if you did miss something, you still have a few minutes.

An ideal last question, always, is this one: "Is there anything I should have asked you and didn't?"

Leave no loose ends: Establish some way to ask more questions (e-mail is ideal), and if you may need another appointment, make it while you are there. You can always cancel the slot you have, but it's unsafe to count on getting time at short notice.

Before you go to bed that night, review and flesh out your notes. No matter how well you take notes, there is never time to write down all the details—so do it now, while you still remember. Also, jot down any questions that came to mind as you were on the way home. In an ideal world, you would even type your notes in a full version. At the least, review them.

As you review, put red asterisks (or some other prominent mark) beside the bits that really get your attention; probably they will need to appear (in some form) in the final article. You are standing in for the reader, remember? What intrigues you now will almost certainly intrigue the reader. Let your less-informed, just-starting-the-research self leave a trail for your later, know-it-all self.

And for such conscientious virtue, you get a bonus: By reviewing the material on the same day, you move it from short-term memory toward long-term storage—where it needs to be when you sit down to write.

Do not regale your friends with everything you learned. If all went well, you came home bursting with great stuff. You are just dying to tell—*and you need to feel that way when you write.* Do not squander all that good steam on a dinner party. Do not talk socially about a story till you have finished writing.

This peculiar advice is born of something I saw happen again and again: The writer would do the first few interviews and come back all excited to tell me how X said this, and Y said that, and did I know that such-and-so? ... She'd go away to write, and I'd be waiting happily for this dynamite story. But when the manuscript arrived, all the fascinating facts, stories, and quotes, the cream of the story, would be conspicuous by their absence. I'd say, "What happened to the material about such-and-so?" The writer would look puzzled; then the memory would slowly return, as if it were trudging in from some polar zone of the mind. "What? ... Hmm. Oh ... Oh! I'd forgotten! Yes, that *was* a good example, all my friends loved that story ... "

About the fifth time I heard "My friends loved that story," I concluded that every socially competent adult must have a brain center that prevents our becoming bores. Call it the MEGO, after the editorial shorthand that stands for "My Eyes Glaze Over." This hypothetical MEGO keeps count, so that once we've told a story several times, the material drops from active memory as too old, not to be told again. No kidding. I think something much like that happens when writers dine out on their interviews.

Brainstorming the material with your editor or a writer friend is okay, however, because not only is it work, it feels like work. Brainstorming seems not to rouse the MEGO, I think because you are not entertaining each other. Nor are you polishing stories in isolation from the rest of the material. You are poking at the stuff together, looking for highlights and unseen connections. The writer should expect to leave such a session all charged up to write.

Between the research and the writing comes a period of immersion, which can last hours, days, weeks, or even months and years (for a book), depending on how hard the material is. You should not write till you understand at least one layer deeper than the piece will go, which means that you have to put in time, often spending hour after hour with little apparent result. You read and ponder; you mark and remark your notes (which come to look quite battered); you come up with new questions and find out the answers; you ponder some more.

A period of wandering in a formless void is necessary to any writing of substance, simply as part of achieving a new synthesis. The void is more intense for science writers, however (or so I believe), because of their frequent need to master new material. Inexperienced writers sometimes abort articles because they cannot stand any more struggle, so they start writing anyway, leaving out the "confusing" parts.

This phase will be less scary once you have been through a big immersion and come successfully out the other side, because then you'll know in your bones that feeling hopeless is just a phase. After that, you'll probably find the process rather fun. It has all the joys of solving the New York Times crossword puzzle, plus you are being paid to satisfy your Elephant's Child curiosity.

A few suggestions for learning what you need to know: Try starting with the hardest parts, as my piano teacher used to advise. Then the rest will seem simple by comparison.

Treat immersion like cramming for a test: zero in on the skeleton, the few central words and concepts, which you will spot because they keep coming up again and again. They'd be on the test because they're important, and they belong in your mind (and probably your writing) for the same reason. Using a good encyclopedia of science and technology, look up every word you run into *more than once*.

Emphasis, *more than once*. If you try to look up everything, you'll be overwhelmed. I've seen it happen.

Don't get derailed by some tangential detail, like the reference to the Elephant's Child a few paragraphs back. Did you stop reading to puzzle over it or, worse yet, to look it up, and if so why? Either you grew up on Kipling's *Just So Stories* and you understand precisely how excessive that curiosity was, or you know from context that it must be some kind of big curiosity. Either way is fine. You can tell the phrase is a throwaway because if you remove it from the sentence, nothing happens. There's no loss of meaning, only of a chuckle for some few readers.

Curses on the many teachers who seem to have taught a whole generation to stop dead until they understand each and every word. If babies tried to learn that way, none of us could talk. Go for the big picture. Trust the context.

To shortcut immersion in the future, start building a wide base of miscellaneous information. The less scientific background you have, the more you may wish to subscribe to *Science News* and the *New Scientist*, both of which offer what you need: an ongoing education in science and technology, including the latest news. Both are written accurately at a mostly lay level, and they will arrive at your door each week. If you subscribe, you can use their web sites, and if you save the magazines, after a while you'll have research material on almost anything you might ever need to figure out.

Don't work at it. Just read for pleasure, and you'll get a lot by osmosis.

Science, the journal of the American Association for the Advancement of Science, is also weekly, and its front section of science news is without peer. Because it is a scientific journal, however, and priced accordingly, you may decide to read it at your local library.

If you like science but do not enjoy an intellectual tussle, even in retrospect, you may want to become a specialist, writing in one area only. In that way, you will shorten or eliminate the period of immersion.

Writing
Getting Started and the Structure

> I do not always
> love to write. I love
> having written.
>
> —*Anonymous*

Now let's go on to actual writing, which we will discuss
in terms of articles. While you may eventually write
books, most books develop from articles. So:

You've got your story idea, you've done good interviews, and you've studied the background material. You
know who, what, where, why, when, how, who paid, and
why it matters—or you think you do. Now what?

**In writing science, there are three main criteria—lu-
cidity, lucidity, and lucidity.** Once the train of thought is
crystal clear, every fact in place, charm and flow may well
emerge. Neither helps, however, if the facts are not
straight. Indeed, premature charm can get in the way, because you will want to leave it in. That is why you should
... go to the next maxim.

Begin writing by not writing: THINK. I seem to hear
you saying, "*What?* Think some more, after all that immersion?" Yes, because now you need to think about how
you should present the material, a subtly different question—though in practice, of course, the two phases of
thinking blur and join.

If you start to write without knowing what you want to
say, you will have to write multiple drafts—a painful
process, even its practitioners would agree. On the other
hand, if you have thought to the point of boredom, you
could be writing in regurgitation mode, which is dull for
you and dull for readers. Best, if you can get there, is the
middle road: Start to write when you're clear enough that
you won't go wrong—but are still thinking, still excited,
still able to be surprised as the last few details click into

place. If you think while you write, you will enjoy it more
and your prose will be more muscular and engaging.

If you have ever painted a room, someone surely told you
the old painter's adage: It takes more time to get ready than
to paint. And you surely found it true: patch the holes in the
plaster, sand the woodwork, sand the patches, vacuum up all
that fine grit, find a hole you missed, patch the hole, vacuum
up all that fine grit, put masking tape along the window glass,
go buy more tape, tape more windows ... By the time you lay
down drop cloths and discover that you have only semigloss,
it's time for dinner and you haven't even wet a paintbrush.

Yes, but the next morning you do the whole job in three
hours, and there's no need to razor the windows or scrub
paint off the floor. And it's the same way with writing.

**Think about the readers, emphasis readers, plural. Read-
ers come in clusters.** There is never only one, though one
will be central. When you write, you will address that key
reader directly, thereby rousing your social skills. The other
readers will listen in and benefit from the occasional aside
(or joke, or whatever) that you tuck in for their benefit.

When I write, I can almost see my readers, a ghostly
crowd installed in my head by the city editor of the *Corning
Leader*, the newspaper of Corning, New York. I was the new,
young "wire editor," hired to sort through the national and
international news that the Associated Press still sent by wire,
then decide what news to put on the front page (and only
the front page, because local news and the grocery pages
were why people bought the paper). The city editor was
giving me a one-minute training.

News included the doings of President Nixon and the war
in Vietnam, of course, and I was to remember that some
people got *all* their news from the paper. I was to squeeze in
all the highlights.

But world events were just a start. "People like to laugh,"
said Walt. If something amusing came through, I was to
use it.

And "Most of our circulation is on the rural routes," said
Walt. "They're farmers. So anything that matters to agricul-
ture, that's news."

Then there were the vineyards at the nearby Finger Lakes.
"Because we have vineyards, we have Italians," said Walt.
"And they're Catholic. So anything the pope does, *that's* news."

And, finally, it was winter. February, as I recall. "At this time of year," said Walt, "people get tired of snow. Now, you'll notice that the AP sends through a picture of girls on the beach in Australia almost every day. Save the good ones. I want to see one of those pictures on the front page about once a week."

After that masterly briefing, Walt and I got on pretty well, except the day when I put the death of Janis Joplin on the front page, top left. The sordid death of some rock star was *not* news! screamed Walt, scarlet with rage. "It is to our *young* readers," I mumbled.

So: one front page, many readers. One book or article, many readers. One magazine, many readers.

From that viewpoint, your goal in writing is to capture and serve as many different readers as possible, yet stay focused on the core concern shared by the subgroups. You directly address the key reader, offering 100 percent of what that person needs (e.g., rudimentary world news). Then you throw the others a bone whenever one comes to hand (corn blight in Ohio, Pope visits Venezuela, Janis Joplin dies). I am going to discuss one example in some detail, in the hopes of infecting you with a good feel for how to proceed.

Identifying the core reader is a crucial decision, one that needs thought. Suppose you have an assignment to write about infant vaccination for the *Johns Hopkins Magazine*, whose readers are Hopkins alumni. Most have advanced degrees, many are medical or scientific professionals, and those older than forty are predominantly male. As a whole, do they have any personal *need* for information on vaccination? No. The group includes young parents, but they do not expect pediatric advice in their alumni magazine. It's similar for alumni physicians: they go to journals and conferences for their medical updates. Anyway, both parents and docs need detail that you could not sensibly inflict on other readers.

So your article might best address the educated curious— who I would argue constitute the core readership for almost *all* science writing (other than self-help). But only the very curious, such as the Elephant's Child (who wanted to know what the crocodile ate for dinner), are curious about anything and everything. Most people care only about things that affect *or might affect* the world they live in.

On that basis, try visualizing the core reader as someone—any sex, any age, anywhere—who might have an effect on public health policy: a senator's aide, perhaps, or the senator herself. If you offer this person a solid translation of the new scientific material that a policymaker ought to know, the article will automatically call up most people's sense of might-affect-me. It will be relevant.

The key issue, in this case, is herd immunity to disease: the idea that if almost all of the group are immune, even the unvaccinated few are safe because they never get exposed—how could they? Everyone else has been vaccinated.

If you can deliver that concept so fast that the reader gets it at a glance, your policy wonk will stop flipping pages—ah yes, this could be important. So, *how much* of the herd must be immune, under what circumstances? Some babies react badly to some vaccinations, he has heard—what percentage? Of which vaccinations? (Make sure you find the latest figures.) How do those numbers stack up against babies saved from major infectious disease? Is the U.S. population as a whole sufficiently vaccinated against infectious disease? Is there any good argument *not* to vaccinate? Does the possibility of biological terrorism change the picture? You can see how the article would develop.

For a stark contrast, now suppose you have an assignment from *Family Circle* to write advice about vaccinating baby. *Family Circle* amounts to a professional journal for women building homes and families, so its readership is relatively narrow. You can bet that almost everyone who reads the article will be a young mother who wants to become a better mother: your key reader.

Now step into that woman's shoes. Ask yourself, If I were a young mother with a new baby, what would I want to know about vaccination? What would I already know? What misinformation might I have? What might I be afraid of? How realistic is that fear? What *must* I know, to make good decisions? Your article will need to address all those issues, in language accessible to the reading public at large—let's say, a person with a high school education (but smart, never forget—she's *smart*).

This woman may have an interest in public policy, but her overriding concern is how to do the best for her own baby. Obviously, your discussion of herd immunity will be radically different from the piece you wrote for Hopkins; you'll be spelling out the trade-offs in terms of individual risk. In addition, the key reader will have a more sophisticated

sister, who may only skim your article because she read
about herd immunity some years back. So she thinks she
knows ... but you've found some new and important information about particular vaccines. Can you think of some way
to highlight the new stuff for a skimmer? (Yes, put it in a
box.) She may have more sophisticated questions, too. For
example, Does it help to postpone some but not all of the
usual infant vaccinations? If so, which ones? If the mother is
nursing the baby, does that make a difference? Not all the
readers would have thought to ask those questions, but they
will all care about the answers.

Then consider the pregnant reader. Since nursing helps the
baby, she needs to know it now, while she's deciding
whether to nurse. Is there anything else she needs to know?

Even more important, imagine a young mother who reads
with difficulty. She is struggling to read your article only because she is worried—a neighbor said vaccination might
hurt her baby, and then she happened to see the cover blurb
in the grocery store. That final reader constitutes a small but
critical audience: critical because she and others like her do
not normally read magazines and newspapers. Basically, your
article may be her only source of information. For this
reader, you might take special care to state each key point
simply, perhaps in boldface, before you elaborate.

So far as I know, no such pair of articles exists, though
herd immunity does. I picked Family Circle because it is an excellent source of basic medical information, translated out of
medical jargon yet helpfully specific. The editors serve their
readers well.

The take-home lesson is that, for any topic and any publication, you should hold in mind a central reader, along with
a cluster of others who have somewhat different needs and
backgrounds. Whenever you get a writing assignment, ask
the editors to tell you about their various readers.

**As you start to think through a piece, imagine yourself as
each reader in turn. Who are these people? What does each
one need and expect from you?** What will each group want
to know? If you meet one particular reader completely, will
that do most of the job for the others? Yes, that's the primary
one, the reader. Knowing the reader early on will help you
decide how to approach your article, and later it will help
you choose vocabulary, examples, and analogies.

In general, the readers of serious nonfiction are intelligent and curious, but otherwise varied—old, young, male, female, East Coaster, Californian, Midwestern, with and without various levels of education and background. Any given article might even reach a few specialists, checking to see what the general public is reading. (Though tiny, this group is important to you because they will remember your name if you appall them.) At the opposite end of the spectrum falls a group that you might encapsulate as eighth graders gathering clippings for a class project in science—future readers, as it were. Most readers fall somewhere in the middle, having at least some college education. The dominant trait that all share is curiosity.

The difficulty is that "some college education" leaves little common ground. The days are long gone when all undergraduates took a foreign language and four semesters of science. The readers who did take college chemistry may remember little, or wrongly, or what they remember may be outdated. Even people with PhDs in other sciences probably know little to nothing of any particular topic outside their field.

Bottom line: any background information that the readers need you must supply—in such a way that you appear to be only reminding them, or that the better-informed readers can easily skip.

Think about the subject matter and mark your material for use in writing. With your reader(s) held in mind, review all your notes and printed matter so that all is fresh in your mind, seen as a whole. If you are writing a brief news item, such a review may take fifteen minutes. For a major feature, it may take several days. Relax: You will get the time back when you write. Just keep combing through. (You can see why immersion and planning tend to blur.)

Now that you're no longer struggling to understand, the pondering phase is fun, a time when you get the payoff for the asterisks, comments, and questions you left for yourself in your notes. I almost always find unexpected gems in the early interviews, facts and comments that I had forgotten or hadn't known enough to savor at the time.

To write a long piece, you'll want to leave yourself a trail back to any quotes and facts you need, without hours of rummaging. So whenever you find a gem, something you will want to use, write a yellow sticky and place it in the margin, poking out. Soon your books and notes will fairly bristle with stickies saying things like "Feynman paint story" or "Good e.g. protein folding." Later, as you write, you'll be able to move right along, maintaining a forward momentum that will keep the piece lively.

Before you start to write, write a head and subhead— good ones, not perfunctory. The process will force you to get precise about both topic and approach.

As a unit, the heads have two jobs: to lure the readers in and to constitute a fair billing. Consider pheromones, the chemical signals with which animals (including us) attract mates—moth pheromone does nothing for rutting bucks and vice versa. In the same way, the allure of your headline should speak specifically to the right readers, the cluster of people you are talking to. Obviously, the articles for *Hopkins* and *Family Circle* would carry very different heads.

To write headlines, imagine your flock of readers sitting across the desk and ask yourself, "What am I really trying to say?" Then start typing, brainstorming with your keyboard. Don't worry about whether you've got anything good. Just keep at it till you generate a flow. Something like this:

THE MUMMY'S TALE
What mummies tell us about their life and times
What mummies tell us about their lives and times
What a mummy can tell us
Diagnosing from 3,000 years away: a Swiss doctor ...
Was acupuncture used in ancient Europe? Oetzi's tattoos say maybe so. CHECK SPELLING
Was acupuncture used in ancient Europe? A mummy's tattoos say maybe so.
Was acupuncture used in ancient Europe? Mummies cast light on this and other questions.
Acupuncture in ancient Europe? Maybe so. Mummies reveal all.
Acupuncture in ancient Europe? Neanderthal civilization? Mummies can still speak.
Mummies can hint.
Mummies are full of silent hints.
Hints from Mummies
The Mummy's Whisper
The Mummy's Tale

75

Acupuncture in ancient Europe? Neanderthal civilization? Ask a mummy. [EUREKA!?]

Acupuncture in ancient Europe? Neanderthal civilization? ASK A MUMMY.
(Yes, the subhead can sometimes precede the head.)

And so on. Fill several pages.

In this case, I happened to get a workable head right away, so I went straight on to subheads until a too-long subhead produced a new candidate for headline. You can see I was starting to get silly, often a good sign. Once you create a few outrageous heads, you'll be chuckling, and then you may do a double take: Yeah, but actually, there's something in that ...

Brainstorm till you run dry and take a break. When you come back, you'll be able to assess which candidates show signs of life.

It can help to show two or three headline combos to other people—but do not ask them for suggestions unless they know your material. Ask them which headlines make them want to read. For each one that has allure, ask what they expect from the article. Then ask yourself, Is that roughly what I plan to deliver?

If your most alluring heads keep dwelling on some part of the topic you were not planning to emphasize, you might want to reconsider. Your subconscious (i.e., your muse) may be trying to tell you something.

If it's not written down, you don't have a plan. Make a plan, following the advice once given by Alexandre Dumas père for three-act plays: The beginning (first act) should be clear, clear, clear; the middle (second act) should be interesting, interesting, interesting; and the end should be short, short, short. Your written plan may be very simple, especially for something short: head and subhead, idea for the opener, idea for the closer, plus a list of three to five major points you want to make in between. I often write from that little. Others make elaborate outlines.

You will need to experiment and see what works for you—probably some version of however you wrote as a child. Did you make detailed outlines, as the eighth-grade teacher said to do, and it worked? Or did you write, then produce the outline later, since the teacher wanted to see one? Or were you somewhere in the middle? Maybe you outlined the main points before you wrote, but filled the rest in later for the teacher?

Wherever you stood in that spectrum, you are the same person now, albeit using bigger words. So the same approach will likely be congenial.

If making and using detailed outlines is your natural way, I suggest you think through organic structures as discussed in the next few pages. Make an outline that will execute the shape you found inherent in your material, with special attention to the opener and closer. Then write.

If you are more of the intuitionist school, you can skip the outline, but do write the heads and list your three to five major points. Then plunge right in: write the opener.

The opener: Imagine your readers, ask yourself what you want to tell them, and start typing. Keep taking a run at it, brainstorming with your keyboard, much as you saw me doing with heads and subheads, until suddenly something feels right. The approach just ... fits. You know it's a Yes even before you see why or where to go next, and the imaginary reader(s) in your mind will be nodding, eager to hear what comes next.

An opener should follow seamlessly from the head. It should rivet the reader, establish rapport between reader and writer, and strike into the heart of the story—all in a single paragraph, if the piece is very short, say fewer than five hundred words. For a New Yorker–style piece, as many as six paragraphs will be okay, but the first paragraph still has to be riveting. Aristotle's advice, to begin in medias res (in the middle of the action), has been helping writers for two thousand years now, and it will help you, too.

Some people write the body of the text first and save the opener to write last, to benefit from all the clarity that the writing brought. If that works for you, great.

Others find (and I am one of them) that by not going on till we get the opener right, we gain a clarity that shows up in the whole rest of the piece. For me, the opener is a sub-structure, a footing that must be in place before it feels safe to erect the building. Since I began doing it that way, I spend more time writing the opener, less time writing the body of the piece, and less time overall.

It will help you to have the closer in mind as you write the opener, for though you may later change your mind, you must still embark in a specific direction.

The best openers are extremely specific, even concrete. They grab hard and then move immediately into exposition, four to six paragraphs worth, that is above all *clear*. Consider this first paragraph from "Cooling the Lava" in *The Control of Nature* (Ferrar Strauss Giroux, 1989) by John McPhee:

> Cooling the lava was Thorbjorn's idea. He meant to stop the lava. That such a feat had not been tried, let alone accomplished, in the known history of the world did not burden Thorbjorn, who had reason to believe it could be done.

He meant to stop the lava. Audacious!!—and totally clear. Already, I am well and truly hooked. Now you'd have to bore me for several *pages* before I'd quit, so McPhee has plenty of time to exposit Iceland's economic dependence on Heimaey, the country's only harbor, and to describe Thorbjorn's early efforts when an emerging volcano threatened to fill that harbor. But imagine if the piece had started with the third paragraph: "Heimaey is pronounced 'hay may.' Vestmannaeyjar is more or less pronounced 'vestman air.' The town on Heimaey is the only place in the archipelago inhabited by human beings, ... "

Here's another opener, by Cullen Murphy, from "Lulu, Queen of the Desert" (*The Best American Science and Nature Writing*, 2000, reprinted from the *Atlantic Monthly*), which relies largely on a fillip of surprise, even exoticism:

> Julian Skidmore is lithe and petite, with small wrists and delicate features, and a serene but determined countenance. Watching Skidmore at work for a while, her auburn hair held back by a blue ribbon, a glint of light catching the small pearl in each earlobe, I was reminded of Gainsborough's portrait of the young Georgiana, Duchess of Devonshire. Then Skidmore removed her left arm from a camel's rectum, peeled off a shoulder-length Krause-Super-Sensitive disposable examination glove, and said, "Can I make you a cup of coffee?" She had completed eight of the morning's sixteen ultrasound scans. It was time for a break.
>
> Skidmore, an Englishwoman known to everyone as Lulu, has emerged during the past few years as among the fore-

most practitioners in one of the world's more improbable growth industries. There are many reasons why *Camelus dromedarius*, the single-humped dromedary camel of Africa, Arabia, and southern Asia, might have deserved to become a focus of scientific investment . . .

The image of Georgiana, Duchess of Devonshire removing her left arm from a camel's rectum is a surprise—admit it.

Reading that first paragraph, I was charmed and curious, as well as faintly tickled to be addressed as a person who, of course, would know this famous painting. (We're all cultured people here, right?) Notice, however, that Murphy has taken no risk: Georgiana is a throwaway. Whether the reader knows the painting or not, one gets enough sense of the painted beauty from the living one to be lulled along.

Notice, too, the shoulder-length Krause-Super-Sensitive disposable examination glove and the eight out of sixteen ultrasound scans. Such delectable detail not only takes me there but also helps me relax as a reader. I feel, Oh, it's okay. I'm in the hands of someone whose eyes are wide, wide open. I can trust this writer.

In the second paragraph, having got our attention and introduced the main character, Murphy moves immediately to lay in the footing—the extreme difficulty of scientific camel breeding, Skidmore's role in that enterprise, and the several reasons for doing it. To race the animals is one. To develop animals for human meat and transportation after the climate warms is another.

That latter reason is left unsaid till the closer, however, also a point worth noticing. There's a hint in the opener, but a hint only. I kept reading along, intrigued by the "baroque masterpiece of biological engineering" that the camel is, until Murphy thumped me over the head at the close.

"We could do a lot of good for other countries where they really do need the camels for meat," Skidmore said. "Where they really do need them for milk. Where they desperately need them for transport. Worldwide, camels are becoming a much more important animal, as we kill off our environment by building everything up and draining the water out and pulling up trees. Before long a lot of the world is going to be desert—the desert is enlarging all the time. Camels will be one animal that can survive all this. We'll be farming camels instead of cattle and sheep. At the end of the day they're going to be a lifesaver."

Skidmore laughed. "Global warming," she said, "could be very good for me and my camels."

The idea that smart, commercially savvy people are spending many millions of dollars against the climate warming . . . is impressive. Yet, if the piece had begun with a call of alarm over global warming, even I (who also worry about it) would have flipped the page. People don't want to hear urgent alarm. Don't bother. Charm the reader, as Murphy does with his civilized yet earthy urbanity, and maybe you can not only give your readers an agreeable hour, but also strike a blow for global awareness—as long as you first get their attention.

You will learn a lot about openers if you analyze the opening cadences of any article that you admire. Watch how the artist grabs you fast, then chunks in the background with big slashes of gesso. The tone may be casual, but every brushload hits exactly so. The entire surface is prepared in a few powerful strokes.

In the middle movement of a piece, the pace slows as the matter complexifies (part of what makes it interesting, interesting, interesting). The writer touches in subtleties of color and detail. In the end, often only two to three paragraphs long, one final touch snaps the whole picture into focus in a way that is unmistakably final, as you just saw happen with Lulu Skidmore.

Even if the article as a whole is not a narrative, consider including a brief history in your exposition, because a technology or a scientific question is often most clear at its inception. For example, here is Malcolm Gladwell (author of The Tipping Point) describing the invention of television in the New Yorker (May 27, 2002, p. 112):

The idea of television arose from two fundamental discoveries. The first was photoconductivity. In 1872, Joseph May and Willoughby Smith discovered that the electrical resistance of certain metals varied according to their exposure to light. And, since everyone knew how to transmit electricity from one place to another, it made sense that images could be transmitted as well. The second discovery was what is called visual persistence. In 1880, the French engi-

neer Maurice LeBlanc pointed out that, because the human
eye retains an image for about a tenth of a second, if you
wanted to transmit a picture you didn't have to send it all
at once. You could scan it, one line at a time, and, as long
as you put all those lines back together at the other end
within that fraction of a second, the human eye would be
fooled into thinking that it was seeing a complete picture.
The hard part was figuring out how to do the scanning . . .

The historic approach allows you to cater to the full range of
readers—an elementary explanation for English majors, an
interesting history for those who already understood the
technology.

**If such a thing is possible, the closer is even more impor-
tant than the opener, because it governs the reader's last
impression.** Readers should come to the end with a pleasant
sense of completion, as if dawdling over dessert at the end of
a meal.

For that reason, resist any urge to write a grand, booming
conclusion, suitable for declaiming from a pulpit—like the
one I wrote as a college freshman: "John Brown's trial is a
blot on the American escutcheon." (What *is* an escutcheon?
I'm sure I didn't know then, either.)

If the urge to boom strikes you, take two aspirin and sleep
it off. In the morning, emulate Cullen Murphy:

**Structure your piece in such a way that, when your train
of thought comes to an end, its caboose just happens—of
course not, but it should *feel* that way, natural and inevitable—to be a good place to leave the reader. That place
might be a scene, a new insight, a question, or simply a
final image that encapsulates the major idea.** Often, as in
Murphy's piece, the conclusion enlarges the picture (oh! It's
about more than racing!), and it may well bear on the
reader's eternal question, why anyone should care.

The caboose must also be obvious *as* a caboose. It is frustrating for a reader to turn the page expecting more but
finding that no, it's all over. Even if you must leave your
readers in an ambiguous frame of mind, because ambiguity
is the truth of the matter, do it cleanly. Make your good-bye
unmistakable.

If the story has mutated under your hands, you may not

always know you are writing the closer, or at least that often happens to me. I'm following the string in my own mind, writing away till I notice I have nothing more to say. Yet my planned conclusion feels dull and unconnected. So I look through my papers and ... nothing. There's nothing here that wants to be added. I scratch my head, I read the draft again, and I have a eureka moment—hey! I *have* a closer! I just need to touch it up so it feels final!

The closer as surprise must be common. As an editor, I can't tell you how often I have dropped someone's last two booming pages, then massaged the perfectly good closer the author already had, if he'd only noticed.

Those writers, like me, probably had a high school teacher who taught the tell'em, tell'em, tell'em method: Tell 'em what you're gonna tell 'em (opener), then tell 'em (the body of the piece), then tell 'em what you told 'em (the closer). That antique advice can still work for speeches, if the subject is so complex that listeners need the repetition. It is deadly in the written word, however, not least because it generates those booming conclusions. In his heart of hearts, the hapless writer knows that the final repetition is boring, so he amps it up to ludicrous.

Modified Dumas père works better and is so important that I'll say it yet again: The opener should be clear, clear, clear; the body should be interesting, interesting, interesting; and the closer should be short, short, short.

Within the general framework of get in (clear), tell 'em (interesting), and get out (short) lie a thousand possibilities, each of which has a particular organic shape. As you go along, try to "see" that inherent shape in the material itself—a spiral, meander, beech leaf, delta, or such—then use it to structure your article. In this way, subject and packaging will have the same shape, and any structural problems will be small ones.

This approach also enhances rapport with the reader because the organic brain knows organic shapes. Both writer and reader have been living with these forms for all the years of our lives, and they are deeply, deeply familiar. As a result, an article having such a shape holds writer and reader on the same wavelength. Both parties know where we are intuitively —an advantage so big that you should maintain it at any

cost.

When you first try to see an organic shape in a pile of
notes or a sketchy draft, even when you get an image, you
may feel you are making it up. Or it may feel like trying to
navigate on peripheral vision. Do not worry. Even for me
(who came up with this idea), and even after all these years,
it is often *as if* I see a shape. But whatever I "see," I can trust
it, and I think you can, too. You have nothing to lose by giving this approach a try.

It helps to ask yourself how the material "wants" to be.
You will often find several clumps of stuff that clearly belong
together, much as a molecule forms when its atoms share
electrons. Then you can look to see how the various units of
thought attract and repel one another to form a larger shape,
which will suddenly look familiar.

If one shape is not working, try another. The right one
will naturally accommodate all the important material, and it
will also display a clear, compelling way in and way out.

Many stories are spiral, a common shape and one that only
living things can produce. For example—I am going to use a
clichéd story to make sure that we are all visualizing the
same one—take the classic sick toddler story:

Open on a scene in the doctor's office, where Sick Tot is
being examined by Concerned Clinician. The tot is pale and
listless and spends the whole visit in the lap of Loving
Mother. As well as bringing on the main cast of characters all
in one go, this approach gives the writer a natural opportunity to explain the kid's Devastating Ailment (and all of it's
clear, clear, clear). The story then jumps to the lab, to explain
some Wonderful Research that has led to a new treatment for
which the cute kid enters the trials. Maybe now we have a
few ups and downs, and maybe we meet some other kids
and parents, and we learn more about the disease (interesting, interesting, interesting). In the closer, we are once again
in the doctor's office but now the cute kid is up and explor-
ing, while the loving mother and the caring researcher are
gratified. The mother looks ten years younger. The devastating disease is better understood, and help—some help, anyway—is on the way for other cute tots.

There is nothing wrong with such stories, by the way. The
reason they have become clichéd is that they work so well!
Writers have worked that format to death, and even *so* the
stories mostly work. I guess we all care about sick kids. It

helps that the reader knows, uneasily, that this story is non-fiction, so that the happy ending is never guaranteed.

Did you see the spiral? Or perhaps, more exactly, a helix? The story loops, so that the reader enters and leaves the narrative at the same place (imagine the doctor's office as one o'clock on the circle). But the second time around we have traversed time and are one loop down: we understand things more deeply. Or you could imagine the story spiraling upward, from very sick kid to almost well kid. A good writer can produce both effects at the same time.

I put a spotlight on the structure as I told that story, and no doubt you found it crude. A nonwriter reading the actual article, however, would not hear the clockwork grinding. She would be wondering how well the new treatment worked. After all, were you aware that Cullen Murphy began and ended in the same place, at work with Lulu Skidmore? Returning to the starting point, with a difference, frames the events of a story to produce a sense of homecoming and completion.

Let's watch a class act pull it off: Peter Matthiessen, no less, in "The Island at the End of the Earth," originally printed in *Audubon*. I found it in the annual *Best American Science and Nature Writing* for 2000. As usual, the gentleman is on a quest, this time for South Georgia, an icy island most of the way to Antarctica, yet with an abundance of wildlife—still, though the whales are gone. As the tourship leaves for South Georgia, Matthiessen opens with an incantation:

> The ship sails from Ushuaia, Argentina, at 6 P.M., due east down the Beagle Channel. To the north and south, the mountains of Tierra del Fuego are dark, forested, forbidding, showing no light or other sign of habitation. Already, a soft swell tries the bow, and a gray-headed albatross appears out of the east, where high dark coasts open on the ocean horizon and the last sun ray glints on the windy seas of the Drake Passage.

The group reaches the island. The author tells how it used to be, bustling with men who harvested twenty-five whales a day. He mourns the whales. He also visits fur seals, elephant seals, penguins, snow petrels, and a sooty albatross, and he closes where he started as the ship sets forth again.

The ship rounds the rock islets that lie off Cape Disappointment, and the ghostly snows of South Georgia's windward coast come into view. The mountains fade in the starboard mists as the ship bears southwest for Elephant Island, off the Antarctic Peninsula. The bow rises and falls on the long swells of the Scotia Sea, smote by the night wind and the stygian blackness of a fast-moving squall, crossing drowned mountains.

Notice the many deliberate echoes: "The ship sails . . . ," matched by "The ship rounds . . . " In both passages the author evokes mountains (drowned and otherwise), darkness, and the bow rolling on the swells (soft vs. long) as the ship moves out into a black and windy vastness.

Amazingly, the article holds together, even for this nonbirder, without a single obvious grabber. There is no story line, no conflict, no resolution, no puzzle to solve, no nothing. An old man recounts observations of birds and animals on one South Atlantic island, along with some history of the place. I would have thought such an article would be too aimless, like a river's meander (. . . this way, that way, this way, that way . . .).

Yet Matthiessen makes it work, in two instructive ways. One technique, as we have seen, is the symmetry so clear at opening and close. The reader enters and leaves this universe of words at the same place, pulled through the duration of the trip by a well-crafted spiral.

The other factor is a powerful undertow of emotion, which I will discuss in relation to meanders.

Meanders can also structure an article, but they carry a risk: A train of thought that sways back and forth, back and forth, can seem aimless, even meaningless—an authorial high crime. As readers, we get uneasy if we feel the writer has no plan, no reason to tell us this rather than that. We avoid a writer who cannot navigate his own mind, much as we'd avoid a guide in Venice who cannot find Saint Mark's.

In the name of "fair" reporting, news magazines commit many a meandering story. You'll find an object lesson in almost any issue you care to sample, with a train of thought roughly like this: The subject is X. On the one hand, this. On the other hand, that. When thatians say That, thisians reply

This. When thisians say This, thatians say That. Someday, but not yet, we will have an answer.

A meander is even worse in science stories, because in science, one opinion is always better than another, if we only knew which. One guy's informed guess should be held up to the evidence, not only to some other guy's guess, because validity is a question of evidence, not fairness. Yet evidence tends to show up as missing. Maybe some editor thinks it is "old news"?

To see a meander that works, let's go back to Peter Matthiessen, who certainly can navigate his own mind. As well as a helical structure, he also draws upon the implicit archetype of life as a voyage—one that Matthiessen, born in 1927, may see drawing to a close. The bay of South Georgia was once thick with whales, so many that whalers had no need to even leave the bay. Now there are none. An unspoken parallel: A man who once trekked the high Himalayas in search of the snow leopard now cruises with a wildlife tour.

What drew Matthiessen to South Georgia, he says, was a longing to see "the astonishing bounty of life on the shores of those white, icy peaks lost in midocean, where one might have thought no life could exist at all." As he first describes the coast, it glints like Mardi Gras beads, or a pointillist painting, or something seen through tears.

> Below the cliffs are black-rock beaches, and here the white breasts of king penguins shine against the stones. Nearer, the sun catches the gold ear patches behind the eye of swimming members of this splendid species, as well as the yellow head tufts of the much smaller macaroni penguins, which surface here and there among the dark, round, shining heads of the Georgian fur seals. Antarctic prions go twisting past in their small scattered companies, like blown confetti, and overhead fly kelp gulls and Antarctic terns—the first coastal species seen since the ship left Tierra del Fuego.

Life renews itself on South Georgia, in a great din of roaring, barking seals and peeping penguins, many of them in molt,

> including the big fluffy brown chicks. These stand disconsolate, peeping and chirping in their sweet, rich voices; the parents distinguish them by voice, not by appearance. They trudge along after the adults, bills pointed down, eyes to

the gravel, in a manner that says, "Well, this isn't much fun!" So fluffy are they that their short tail is scarcely visible; they look as if they cannot quite lower their wings ... hopelessly foolish and appealing. As yet, they neither dive nor swim well, and in the salt water they may be preyed on by skuas and giant petrels, which harass and peck them until they are exhausted.

So much life, all of it noisy (the seals "blart"), all of it beautiful, all in danger of some kind, and the great ones are gone. A couple living on the island tell Matthiessen that, in five years, they have seen only two whales from the shore. He keeps coming back to the subject. "Today the Antarctic Ocean is an international whale sanctuary, now that all its large whales are gone."

At first, I thought I was mistaken, even morbid, in linking the declining arc of the whales with the declining arc of the writer's life. After all, the only direct reference Matthiessen makes to his age is to look backward as the ship departs South Georgia "a little wistfully ... , knowing how unlikely it is that I shall return in this life to this remote and magnificent island."

Then I considered the essay's title: "The Island at the End of the Earth." Hmm. In many seagoing cultures, the islands at the end of the world are where the dead and dying go. And when I looked closely at the last three paragraphs, in which the arc of the day also declines, my doubts vanished. Here Matthiessen writes of "ghostly snows," and "stygian blackness" (in classical mythology, the Styx is the black river of Hell that the dead must cross), and snowy petrels that "come and go like lost white spirits between high, dark walls ... In the growing darkness, I climb to the bridge to see Cape Disappointment at the dark end of this mighty island, ringed by explosions of white surf. There is no beacon, nor any sign of man."

Have you ever noticed at the last anything—the last night at camp, the last night of vacation, the last glimpse you got of someone you love—how much that lastness intensifies the experience? What we are about to lose, we experience avidly, no longer taking it for granted. As a great meal ends, we taste our coffee and fruit with special care, so as not to miss a morsel of the pleasure that is ending. Enjoyment blends with loss until the two become almost the same thing.

So it is with South Georgia. Peter Matthiessen has shown us a world intensified, as seen by someone who faces the loss of life and therefore sees its full preciousness. The recurring language of death and loss, even for readers who do not consciously notice it, keeps us oriented to an emotional undertow so strong that it no longer matters how the nominal focus meanders. Penguins, elephant seals, the bravery of Ernest Shackleton—reading, we feel that Yes, it's all one thing, something precious and too soon lost.

Novice writers, if you keep working at it for forty years, you too may be able to write that well. Notice how much less effective it would have been had Matthiessen written something explicit, such as "Life is like a voyage," or "I realized I may die sometime in the next few years, and I wanted to go on one last trip." Readers are intelligent. They only need hints.

The structural lesson is that meanders always need at least one helper, something to contain them. A spiral structure will serve, especially if the piece is short. So will powerful emotion, as in a love letter or certain essays that amount to love letters or grief outpouring. Or a rambling story may be held together by dissonance: hints of hidden action, something important that is happening out of sight. As the writer moves along, the hidden grows more and more apparent till in the end it emerges as the unseen controller, benign or otherwise. You will often see that structure in personal essays, which tend to meander because life meanders.

Background and scientific explanation can depend from a narrative like seedpods hanging from a branch—often an ideal structure. Stories are a powerful way to write science. You tell the story chronologically, moving out along the stem. As events develop, you periodically attach a pod, explaining the science behind what is happening. When the event ends, so does the writing, once you've wrapped up loose ends (the growing tip, which seldom has a pod).

In nature, pods often get smaller and intervals shorter as the end approaches. The same should hold for your article, because by that point the characters are familiar, all the basic explanations in place.

Seedpods also occur with some regularity, which you had also better imitate. The opener is effectively a promise to the reader. If you start with story, you are promising story, so you must keep coming back to your protagonists. The reader

is entitled to know what happened next. Never tell a story
(e.g., sick tot) for three paragraphs and then say to yourself,
Okay, now I have them hooked, so I can get on with explaining molecular genetics. You do not have them hooked: you
have them betrayed. In the same spirit, do not attach a single
pod, promising significant explanation, then go for story only.

The first three chapters of *The Perfect Storm* by Sebastian
Junger (HarperCollins, 1997) are a perfect example of a
seedpod structure. Junger has little plot to work with: the
crew of the *Andrea Gail*, none of whom he ever met, make
ready to depart, with some premonitions of disaster, for a
one-month stint swordfishing on the Grand Banks. In chapter 2 they depart, fail to catch enough fish to make money,
and go farther yet, "almost off the fishing charts." By page
119, they have caught forty thousand pounds of fish and are
just off the Continental Shelf, gunning for home. Three big
storms are converging on the Grand Banks.

Do I need to tell you that these six men will not be coming back? They are heading into one of the most extreme
storms of the century.

As a reader, however, I did not even notice how skimpy
the raw story was, because the pods are so fascinating: the
hard-drinking, hard-working life of these fishermen, with
emphasis on the bars of Gloucester, Massachusetts; the economic pressures on swordfishermen; a careful description of
swordfish ("swordfish are not gentle animals") and com-
mercial swordfishing; several set pieces on some spectacular
storms, tragedies, and heroics off Cape Cod; a tour of the *Andrea Gail*, room by room; a discussion of how a boat rights itself after it has turned ... or not; several wodges of coastal
geography, including a careful explanation of why the Grand
Banks suffer some of the worst storms in the world; how the
Andrea Gail had been refitted so she could stay at sea longer—
which made her slightly top-heavy. As the facts accumulate,
the tension ratchets upward.

The fatal nor'easter itself occupies the three central chapters (of nine): The storm hits, the boat founders, and the men
die, a story reconstructed from every source that Junger could
muster—from survivors on other boats, from the Coast Guard,
from data buoys, from people who have nearly drowned at
sea, from what meteorologists know about major storms. For
sixty-six pages he piles fact on fact, a tour de force of reporting. Watch him tell us how the storm began to build:

For the next hour the sea is calm, horribly so. The only sign of what's coming is the wind direction; it shifts restlessly from quadrant to quadrant all afternoon. At four o'clock it's out of the southeast. An hour later it's out of the south-southwest. An hour after that it's backed around to due north. It stays that way for the next hour, and then right around seven o'clock it starts creeping into the northeast. And then it hits.

It's a sheer change; the *Andrea Gail* enters the Sable Island storm the way one might step into a room. The wind is instantly forty knots and parting through the rigging with an unnerving scream ... By eight o'clock the barometric pressure has dropped to 996 millibars and shows no sign of leveling off. That means the storm is continuing to strengthen and create an even greater vacuum at its center. Nature, as everyone knows, abhors a vacuum, and will try to fill it as fast as possible. The waves catch up with the wind speed around eight pm and begin increasing exponentially; they double in size every hour ...

One can imagine Billy standing at the helm and gripping the wheel with the force and stance one might use to carry a cinder block. It would be a confused sea, mountains of water converging, diverging, piling up on themselves from every direction. A boat's motion can be thought of as the instantaneous integration of every force acting upon it in a given moment, and the motion of a boat in a storm is so chaotic as to be almost without pattern. Billy would just keep his bow pointed into the worst of it and hope he doesn't get blindsided by a freak wave.

And that was the *start* of the storm.

In the same matter-of-fact tone, Junger spends the last nine pages of chapter 6 explaining precisely what happens physiologically in a person who drowns at sea, step by step. Billy Tyne, Alfred Pierre, David Sullivan, Bugsy Moran, Dale Murphy, and Bobby Shatford are surely dead. The reader sees them clearly, sinking down and down and down, limp and open-eyed.

The book continues, however, as it had to. I could diagram its powerful symmetry just like a Jane Austen novel (though I won't). As a whole, *The Perfect Storm* has the shape of a rogue wave—up up up up UP ... HIGHER YET ... IT APPEARS TO HANG ... and finally crashes, to be followed by successively smaller

waves, the events of the last three chapters—the struggle for
life on other boats caught out, as well as the heroic deeds
(including a death) of the Coast Guard rescue teams. The
book ends with a one-page wavelet: one more storm and a
final set of deaths. The ten-page afterword constitutes the
growing tip, no pods: the survivors get on with their lives.

**Organic shapes having some natural beginning and end
provide the best structures, notably helices and waves.
When you write, use the whole shape,** or several, as
needed, and do not be limited by the few I have discussed.
There are many more.

For example, the watershed pattern: you see a stream of
evidence here, another there, a third and a fourth some other
place; they grow, they become rivers, and they ... converge!
The piece ends with something new and big, a veritable
Mississippi.

At the other end of the river, we find the delta. A huge
scientific insight has been developed, perhaps into new
technologies, whole new fields of research, or both. It is
leaving deposits, spreading out and out and out till it meets
the open water.

(Note in passing: I originally wrote not Mississippi but
Susquehanna, a natural thought for an Easterner. Then I
thought, No, some people from other parts of the country
will not know how big the Susquehanna is. I'd better take
the big river we *all* know. Such small but critical emendations will come naturally once you know in your bones, at
every moment, that "the reader" is many and various.)

A book or article will feel wrong if a chunk of the shape is
missing. Do you think Sebastian Junger should have stopped
his book in midstorm, just because the *Andrea Gail* sank? No.
Of course not. The wave of events and emotion had to play
itself out. Similarly with South Georgia: we had to leave, just
as we came.

Ask researchers what shape *they* think their material has—
many will know, or at least have a hunch.

**It will help you to draw your structures, not just to think
them in linear form.** Drawing calls on a different part of the
brain, and you might as well use *all* your tools.

Label the parts. The more complex the structure, the more
you need to see it whole.

If you are steeped in music, you might also try modeling your work on musical forms. All of us, when we read our writing out loud, are calling musical intuitions to our aid. As a friend said the other day, "Oh yes, this is monotonous. You can just tell it has to swell about here."

Some of you could probably take that kind of musical knowing to a different level. Can you think of certain pieces as ballads, others as sonatas or grand opera? Could you see a transition as moving from the key of C to E minor? If your mind works that way, try it.

Since the organic shape is in the material, not imposed by you, it carries information: It can tell you when and where something is missing. For example, do you see a Mississippi with only two tributaries? Such a thing cannot be: therefore, a big chunk of information is missing, if not two or three. Poke around. Maybe you've overlooked something important.

The same is true of a subject's texture. As you probably know, much of nature is fractal (i.e., self-similar), meaning the same at every magnitude. The tiny ins and outs on a sandy beach echo the bigger ins and outs of the rocks in the inlet, which in turn echo the jagged coastline as a whole. You will see such fractal patterns in many things produced by natural processes: they are a hallmark of living things.

So, if information is coming at you leaf by leaf until you are suddenly offered a naked branch, some kind of big, bald general statement, ask for the missing leaves and twigs. Or perhaps what you really want is more branches? Maybe the place you started is just one little piece of something huge and wonderful?

Limber up your mind. When you go for a walk, look around in nature for shapes that remind you of some idea or event. When you read a terrific piece of writing, diagram it. Look for the shape (or shapes).

If the grand finale of your article is clear to you but the structure is not, try a more "logical" approach: Start your plan from the end. Whatever your grand finale, back up. Ask yourself, what does the reader need to know to really get this final part? Good. That material belongs in the penultimate section. And so on. Just keep backing up toward your opener till you get there (or perhaps to a better one yet). A shape should then be apparent—or at any rate present, even if you don't yet see it.

Even when you already have the shape, thinking backward can help refine your train of thought. If the overall structure will be stem and pods, for example, working backward would be a good way to order the various pods. In practice, some events and pods must go together, while other pods can go anywhere, or anywhere before a certain point, or anywhere before some other pod. Flexibility helps, and so does logic.

Writing
The Nitty Gritty

As you write, keep your eye on the ball. I borrowed the sporty image in this mixaphor, hackneyed though it is, because in sports we all know it's true (which is how it got hackneyed). It is hard enough to hit a tennis ball streaking toward you at 118 miles an hour. It cannot be done by a person who is preoccupied with losing, or his appearance, or anything else.

The athlete must stay focused on the job at hand, and so must you: Keep your mind on the subject matter. Think straight, knowing what you want to say and to whom, and say it as clearly and concretely as you can. The rest will follow.

Your initial effort needs to be more or less continuous, meaning day after day, as in all the arts. An artist friend often quotes her painting teacher on that subject: "You must ... go ... to the studio," her teacher would say, slowly and with emphasis. "Once you are there, you might spend all morning sweeping the floor. That doesn't matter. What matters is that you must ... go ... to the studio." Yes, master, I hear you.

For writers, what's more, the time has to be spent actually grappling with the material. You must actively puzzle at it, as opposed to looking at it with despair and wishing you understood or wishing you saw the opener. Your subconscious synthesizing powers cannot get to work until you give them something to work on.

"What am I really trying to say?" is a near-magic question. It will help you get started each day, and it will solve many of the classic writing struggles. When a passage won't budge no matter what you try, stop fiddling. Look away or close your eyes, imagine the reader sitting there, and ask yourself what you want to tell him. Quite often, you'll find you don't know or that you are worried about the reader's reaction. Other times, you'll find that you do know but had somehow gotten married to a sentence or paragraph (or several) that were too elaborate, or that took you in the wrong direction. So—What are you really trying to say?

When the answer comes, write it down as simply as it came. The result will be far better than the version you were struggling with.

If you fear that the readers may misunderstand, disapprove, or be bored, ask yourself why. Is there some fact or idea that you need to put in place earlier, to lay the groundwork for this present paragraph? Are you getting windy and your subconscious knew it?

Write out loud, mumbling or whispering to yourself as you write. Because reading is processed in the speech centers of the brain, any sentence or paragraph that is hard to speak will be hard to read, period. Not a lot harder, of course—but 1 percent improvements have a way of adding up, and this particular habit may be a 2 to 5 percenter.

When in doubt, stop and read the problem passage out loud, actually out loud. Do you feel an impulse to change the wording as you read? You probably should make the change.

For the same reason, avoid lumpy words—words with hard knobs, words that contort your face as you speak them, words that require an effort to spit out each syllable. "Particularly" is an egregious offender. So is "egregious."

Polish your prose late in the process rather than early. The more you work on a piece, the deeper it burrows into your neural pathways, and the harder you will struggle to see it freshly. The more effort you invest, the more every word will seem precious—near impossible to change.

Save yourself some trouble. Write the first draft completely, including examples and technical details as needed, but never polish an early draft. So long as the train of thought is clear, you can leave things a little fluid and keep chugging. Your subconscious is probably doing work that has yet to surface into conscious thought, so that if you wait, some "problems" will have solved themselves. They will seem to evaporate.

If I am moving on though unhappy with a passage, I leave

myself a note IN ALL CAPS, LIKE THIS about what bothered me. Quite often, when I come back to it, I read the note with incredulity. ("I was worried about *what?* This is fine!") Or sometimes the passage looks worse than it did, but no matter—I also see how to fix it. In every case, it helps to approach your problems freshly.

Consider starting a bone heap, a place at the end of the manuscript for discarded sentences and paragraphs that you might yet want—dead examples, for example, or an aside that grew so big it disrupted the train of thought.

The trouble with these items is that one gets *attached* to them, having invested the labor to create them. Hence the value of the bone heap: Knowing you can always retrieve that little gem, you'll find it easy to be ruthless. An example is not quite working? Out!

Occasionally, I do retrieve something, usually a swollen aside that turns out to be something I should have said earlier, a bit of groundwork for the passage where I actually wrote it. The thought had been showing up as missing, and my subconscious got to work.

The true gems will almost always call you back. You'll be starting a paragraph and think … gee, didn't I already write this? Yes you did, and there it is, waiting for you in the bone heap, sometimes in several different versions.

Write with your notes and references open. As a creative person, no matter how well you understand the subject, you need the constraints of genuine facts and quotes. Otherwise, you are likely to improve the stories and ideas past recognition. Use your notes. As a boss of mine used to say, "I don't have time to take shortcuts."

Make sure you put in all your raisins (i.e., fun facts, great quotes, and interesting comparisons). Have you ever eaten a bread pudding that had too many raisins? I can't imagine such a thing, and so it is with writing. You may not be able to turn a brilliant phrase yourself, but if you can recognize brilliant material when you see it, you can come close to a brilliant effect.

I first noticed this phenomenon in editing some articles written by Hugh Kenner, a scholar of English literature and a good friend of Buckminster Fuller's. Kenner turns a mean phrase, but he also borrows beautifully. Here, in an article from the May 1976 issue of the *Johns Hopkins Magazine*, he introduces diffraction gratings, those brilliantly iridescent tools of spectroscopy.

> Midway through the [nineteenth] century, the awesome British polymath Charles Babbage was proud of diffraction-grating weskit buttons. On trips to Europe he always carried one or two more in a hidden pocket, as wonders to buy off savage Italians who might otherwise kidnap a foreign savant for ransom.
>
> A diffraction grating is a square of hard material . . . on which fine parallel lines have been incised.

After explaining the gratings, Kenner moves on to European reactions to the "ruling engine," an American breakthrough in their production, as described in a letter of 1882:

> French physicists muttered "superbe" and "magnifique," while "the Germans spread their palms, & looked as if they wished they had ventral fins & tails to express their sentiments." . . . But causing German physicists to goggle like fish [in case any reader missed the point] was a side effect.

Later in the same article, Kenner moves into the present:

> The apparatus for today's high-energy physics is no more amenable to one-man construction than was the Great Pyramid, and is apt to require a budget of comparable magnitude. So synchrotrons and linear accelerators tend to be one-of-a-kind items, a fortune tied up in each installation, and you scarcely feel entitled to one on the home campus. Instead the pilgrims go where the shrine is, as Periclean Greeks went to Delphi or mediaeval Christians to Jerusalem. Sites for physicists who seek revelation include Chicago, Brookhaven, Stanford.

Take chances. A draft is only a draft—by definition, the right place to experiment. Try writing lushly, or speaking more directly to the readers, or whatever you want to try. You will find the edge of the cliff, the place where you've gone too far, only by going over. Then once you've found the lip, you can stay two paces back.

Be slow to conclude that your experiment was a failure. If in doubt, come back to it tomorrow.

Write using active verbs, just as you were taught in high school English. Sentence by sentence, focus on action (which does what to what) rather than "procedures are" or "the data show that." For example, compare:

The result is a serious drop in blood pressure causing body tissues to not get enough of what they need in terms of blood and oxygen.	Blood pressure then drops, depriving body tissues of the blood and oxygen they need.

Writing in verbs may be taught at every level, but writers should never get complacent about it—especially science writers, because we ingest a steady diet of scientific prose, which will tug us toward writing in nouns.

Be definitely indefinite. Scientists are reluctant to generalize their data, and rightly so. For that reason, any general statements need careful crafting, more than we use in ordinary speech. If your manuscript (in effect the scientist) says "an occasional" case, you should mean *one* case, every once in a long interval. By "a few," you should mean 2 or 3. "A handful" would be 4 to 5. "Several" seems more like 6 to 8, not more, and here we are already in "many" country ... Or are we? It's best to avoid "many" in science writing, unless you truly mean an indeterminate *lot*. Better to pin your scientist down to an estimate like "10 or so," "about 15," "about 20," "some 150," "several thousand," "at least X," "X or more," and so forth.

"Most" is another big offender. In normal speech, we use it to mean anything from "a majority" (could be 52%) to "about two-thirds" to "with rare exceptions." Again, you need to pin the scientist down. If you see "most" in a press release, your index of suspicion should rise up shouting. (And did this press release come from the scientist, a peer-reviewed journal, or the funder?) "Most" can be a weasel word, its big range used to imply more than the science can justify. Don't *you* weasel.

Note also the shades of meaning in expressions like "con-

ceivably," "possibly," "very possibly," "probably," "likely,"
"very likely," and "almost certainly." Does the evidence
"imply," "suggest," "demonstrate," "show," or "prove"?
Deploy such words with care.

**Explain as needed, not sooner and not later, not more and
not less.** If the article's structure is right, the subject will un-
furl like a morning glory, example/case and explanations in-
extricably mingled. Avoid any long patches of bald theory
("First you must understand the uncertainty principle ... ").
Too many readers won't make it through.

Inexperienced science writers tend to overexplain, which
is natural. Photographers love photographs that required
them to wait in the rain for twelve hours, and writers love
explanations that cost them a big intellectual struggle. It's the
hazing principle: If something was hard yet we persisted, we
think it *must* have extra value—as it does, of course. Nothing
you learn is ever wasted.

Your harvest need not appear in the manuscript, however.
Rather, you will often use your new, deeper understanding
to craft an explanation that keeps the idea moving forward
and is true *as far as it goes*. You will become very fond of
phrases like "one of several molecules that do such-and-so."

**If a technical term will come up one time only, silently
translate into something your key reader can get,** like "a
special type of immune cell" or "an icy belt at the outskirts
of the solar system where astronomers believe most comets
form."

In general, unless you are writing as a scientific specialist
to others in the field, translation is always the way to go.
Why say "catalyze" when you can say something active and
specific, like "triggers the [whatever]" or "stimulates the
which to what"? Even the many readers who know what
catalysis is (if they stop to think for a second) will benefit
from the translation. It saves their willingness to concentrate
for any material that really could be tough.

**If you will need a technical term again, as shorthand for
an idea that will return, explain it in passing,** as in this
unassuming little passage by Nathan Seppa in *Science News*
(September 22, 2001, p. 182; I have italicized the parts you
should especially notice):

High blood pressure can lead to kidney problems, particularly in people with diabetes. While scientists don't fully understand the causes of high blood pressure, they know that *a hormone called angiotensin* can contribute to it. Some blood pressure medications offset the *angiotensin's* effects in much of the body, but they aren't as effective in the kidneys. *Part of the problem* lies in the kidneys' unusual design. Blood enters the organs via arteries and then fans out into microscopic capillaries. There, *clusters of cells called glomeruli filter out impurities,* dumping them into the urine. However, the blood doesn't flow directly back into veins heading out of the kidney. Instead, it gathers in another artery and spreads into more capillaries to nourish kidney tissues before it finally exits.

Although the blood pressure medications that have been in use the longest relax the arteries entering the kidneys, they don't always act adequately in the internal kidney capillaries. A bottleneck can ensue that swamps *the glomeruli* with high-pressure blood and damages them, says Barry M. Brenner of Brigham and Women's Hospital in Boston.

a hormone called angiotensin	The focus stays on the easy-to-understand word "hormone," yet slides the proper noun gently into the reader's short-term memory, where it is ready to serve in the next sentence.
clusters of cells called glomeruli filter out impurities	Same device: the blood flows apace, uninterrupted by any definitions, while the reader picks up the new word from context. She will almost certainly comprehend when the glomeruli get swamped in the third paragraph.

Seppa does a nice job of suppressing much knowledge that could have confused the picture: the other causes of high blood pressure, for example; the other part of the kidney problem; the names of the various parts of the kidney's circulatory system; and which blood pressure medications. The unneeded anatomy stays suppressed, while particular medications come into focus several paragraphs later. As the article proceeds, it is as if the selected facts are coated in honey, so that they slide down easy, one pill at a time. No reader will go away thinking, Boy was that turgid, I had to learn a new word just about every paragraph—even though she did.

Build the picture before you supply the name, as in this opener from *Natural History*, by Guy C. Brown (July–August 2000, p. 67):

> The modern cell, now found throughout our bodies, arose a billion years ago from the fusion of two different cell types: big and little. Big ones swallowed little ones but for some reason did not digest them, and the little ones ended up living inside the big ones. Over time, the little ones lost their independence; they handed over most of their DNA and molecular machinery but gained a safe haven within the large cell. The little ones became the mitochondria, and the big ones became modern cells.

The writer gave us a lively mental image on which to paste the word "mitochondria." Simultaneously, he left himself well poised to explain the independent behavior of these important cells.

Start with the question, not the answer, as in another passage from *Natural History* (September 1999, p. 30, by Carl Zimmer):

> Picture a horse in full gallop. Its nostrils flare, its muscles surge, its mane flaps like a flag. A sound track—thundering hooves striking the ground—starts to play in your head. Without those hooves, the horse in this mental movie would quickly slow to a walk. It's their job to hit the ground hard enough to generate a force that can propel the horse forward. On the face of it, however, a hoof seems like just about the worst piece of equipment for the task. A horse's leg ends in what is literally a giant toe (horses descend from an ancestor that had five digits, which evolution has stripped down to only one), and the hoof is a giant toenail that has evolved into a thick wall wrapping around the foot. It's made of keratin, the same kind of pro-

tein found in human nails, and like our nails, it can crack.
We certainly wouldn't try to walk on overgrown toenails.
So how can horses gallop on theirs?

Of course, Zimmer could have started by saying something
like, "The horse's hoof has a complex structure that makes it
very strong. Hoof cells manufacture filaments of keratin, the
same material as in human nails, that are arranged in sheets,
etc., etc." Aren't you glad he didn't?

Rhetorical questions can become an irritating mannerism,
so don't overdo them—but do stay aware that a good question is often the easy way in. We are, after all, members of a
highly curious species.

**Keep the reader with you, joined at the hip, by putting up
a little slalom flag every time your train of thought takes a
swerve or detour.** A word or phrase will do it, of which our
language has hundreds. A few elementary examples:

> For example
> However [not to open a sentence, however]
> Nevertheless
> By [doing whatever you want to call attention to], Jones
>   hoped to ...
> Similarly,
> By extension,
> Some years before,

Also, you can construct the good old "topic sentence" to flag
your turns. Look at this series by Zimmer in an article about
how reptiles and amphibians can stick out their tongues so
very, very far (*Natural History*, October 1999). He starts with
mammal tongues, including the human one. Then,

> But no mammal can compete with the ability of some rep-
> tiles and amphibians to extend their tongues far and fast
> [paragraph on salamander tongues] ... Chameleons can
> hurl their tongues even farther than salamanders, shooting
> them twice the length of their bodies in less than a second
> [by means of a ring of muscle] ... Rings of muscle aren't
> mandatory for sticking out a tongue, however, as demon-
> strated by many species of frogs and toads, ... Kiisa
> Nishikawa, a Northern Arizona University expert on frog

tongues, has been studying a variant of this technique in
the African pig-nosed frog ... The very different tongues
of the pig-nosed frog and the marine toad are suited for
different styles of eating [the quick snap versus slow but
accurate].

And here's the conclusion, just for fun:

Nature, it turns out, has more tongues than the Tower of
Babel.

Note that Zimmer does not call attention to his pun, which
brings me to another rule:

**Never quote anything that passes muster only because it
was a joke.** If you feel compelled to say that the speaker
"quipped" or "twinkled," suppress the quote.

**Avoid "transitions": They are the mark of a structural
problem.** If by "transition" you mean my "slalom flags" or
Zimmer's type of topic sentence, fine. Otherwise, avoid
them: Each train of thought should draw to a close at precisely the point where the next train of thought wants to
begin. If you feel the need for a transition, your train has either gone off on a spur line or stopped short of the station.
Even one sentence short is enough to matter.

**Use quotations from your interviews selectively, weaving
them as highlights into your own well-crafted prose.** You
should quote or paraphrase closely when the words, ideas,
or observations are unique to a particular speaker. (That's another way of saying, Don't plagiarize.) Give credit where
credit is due. What everyone in the field agrees upon, however, you can state in your own way in your own voice, so
long as you get it right.

When you do quote your sources, you may properly clean
up sentence structure, nip out repetition, and even supply or
improve an occasional word (note that I said *occasional*), for
the sake of clarity. If the same idea shows up in several
places, feel free to import the best version into the context
where you need it. You may also mingle two good versions
to make a better one.

You may not, however, alter the meaning by one iota, and

you should preserve the person's characteristic rhythm and vocabulary. Do not turn Mr. Precise ("71 percent of the patients in our last study") into Mr. Folksy ("two-thirds of our patients"). If you supply a missing word for Mr. Precise, make it precise, the one he would have used if he'd had more time to craft his words.

Do not allow the scientist to rewrite the quotes, should you get trapped into showing copy. Ask her to tell you what the problem is, and then *you* devise the improvement (insert the word *probably*, quite often). You must not misrepresent anyone's ideas. On the other hand, you don't want quotes transmogrified into what the scientist might choose to write for his peers, as can happen in Public Relations—which is why press releases often sound unbelievable.

When all quotes are both clear and authentic, human personalities leak into the science and make it extra lively.

Do not write with an authority beyond your personal competence. Science writer Philip Ball, for example, has a degree in chemistry as well as a doctorate in physics. He can write about molecules for pages without crediting anyone else, because he knows.

Nonscientists do better to take it from the scientists, which need not mean writing baby twaddle. It only means that, when you're writing on the forefront, you should get it from the source. Rely on quotes and paraphrases, as in the August 2001 *Scientific American*, where senior writer Wayt Gibbs wrote a short feature about "Cybernetic Cells." Subhead: "The simplest living cell is so complex that supercomputer models may never simulate its behavior perfectly. But even imperfect models could shake the foundations of biology." Without going into the argument—basically, that what we "know" about cells explains very little—let me excerpt fragments to show you Gibbs's skillful patchwork:

> "We're witnessing a grand-scale Kuhnian revolution in biology," avers Bernhard O. Palsson [of San Diego] ... Indeed, reports James E. Bailey [from Zurich], "the cost to discover drugs is actually going up," ... Consider, too, Bailey urges, that geneticists have engineered hundreds of "knockout" strains of bacteria and mice to disable a particular gene. And yet [often] the broken gene causes no apparent abnormality ... "I could draw you a map of all the

components in a cell and put all the proper arrows connecting them," says Alfred G. Gilman, a Nobel Prize–winning biochemist [from Dallas]. But [the components have
no predictive value] ... Bailey compares the confused state
of microbiology with astronomy in the 16th century [be-
fore Kepler proved that Earth's planets move in elliptical
orbits] ...

The authority of a good reporter will take you as deep as
you ever need to go.

**When you're on iffy ground—something you have to dis-
cuss is stigmatized or emotionally loaded—acknowledge
the emotion as common though not invariable, then go forward.** It often helps to write in the tactful third person. For
example,

> Many people find it embarrassing to admit that they are in-
> continent, even to their doctors.
> Many people wonder ...
> Many people are afraid ...
> And so on.

Then move straight along to deliver the antidote:

> People who have this concern often find it helpful to ...
> It helps them to remember that ...
> After a while, these people see that ...
> They are relieved to find that ...

If you are a doctor writing a medical self-help book, you can
point out that you hear similar stories daily—that's very reassuring—and you can speak more directly: "Many patients
tell me X, too often after suffering for months in silence. If
you experience X, you should know that ... "

If you are a nonphysician writing medicine, you can and
should quote the doctors. Direct is good. Many stigmas have
lost power in the last twenty-five years, but not all and not
for everyone. It helps to make full use of the physician's authority and healing aura.

Many people get so locked into their misery that they
never stop to think how many others must suffer the same
way. (The doctor hears lots of people confess that they drink a

quart of scotch a day? He won't tell me that I am a terrible old man? And I can get help without necessarily being sent away? Oh!)

With stigma, the key is to be simultaneously matter-of-fact and nonaccusatory, which is where the third-person format shines. Since each of your readers gets to decide for himself whether he is among any particular "many people," you maintain your rapport with *all* the readers.

Issues of emotion and belief arise outside medicine as well, often as the unbelievable—which also needs careful handling, as in these paragraphs from an article by Lisa Belkin (*New York Times Magazine*, August 11, 2002, p. 34). She is writing about the human tendency to attribute meaning to coincidences, to the fact that amazing things do happen— things that smack of miracles, ghosts, divine intervention, or global conspiracy.

> [A sparrow] happened to appear in one memorial service just as a teenage boy, at the lectern eulogizing his mom, said the word "mother." The tiny bird lighted on the boy's head; then he took it in his hand and set it free.
>
> Something like that has to be more than coincidence, *we protest* [italics mine]. What are the odds? The mathematician will answer that even in the most unbelievable situations, the odds are actually very good. The law of large numbers says that with a large enough denominator—in other words, in a big wide world—stuff will happen, even very weird stuff. "The really unusual day would be one where nothing unusual happens," explains Persi Diaconis, a Stanford statistician who has spent his career collecting and studying examples of coincidence. Given that there are 280 million people in the United States, he says, "280 times a day, a one-in-a-million shot is going to occur."
>
> . . . *When these professors talk, they do so slowly, aware that what they are saying is deeply counterintuitive* [italics mine].

Here Belkin joins the reader ("we protest"), and she spells it out that the experts encounter incredulity all the time. It's expected, not ignorant—a part of the human condition. Her experts address the incredulity directly, so that even those who believe in miracles may keep reading instead of thinking, Oh, this is bosh, and flipping the page.

Form the habit of giving every name, number, and quote a double-check as you type it in, even on your early drafts. It is far easier to get something right to begin with, while you are looking at the material, than to go back and find any particular detail again.

Do not rely on numbers a researcher gave you in conversation. In the fine flow of talking to such a wonderful, receptive, intelligent listener as you, she may well use some approximation that she wouldn't want to see in print. Check the numbers later against published results, which she will have reviewed with care.

Similarly, if you find yourself unsure of any major fact or interpretation, check it out or think your way through it now. If the groundwork of your article will not hold weight, you cannot safely build. The answer is probably somewhere in your notes.

If you think you have to query the scientist, a useful format is this one: "Is it correct to say X, or would it be better to say Y? I am wondering about Z." Doing it this way will force you to articulate your question with such clarity that if you in fact know the answer—as can be the case—you'll realize it. Great! And if you do have to query the scientist, your nice clear question will bring a nice clear answer (maybe a nice quotation, too).

Frequently what you're missing is a chunk of information that is taught to undergraduates in the field, then taken for granted. You've made a great catch.

If you do write a segment from surmise or unaided memory, make sure the item is too small to affect the train of thought. You should also make sure you'll know to check it later. One good method: As when writing notes to yourself, write all doubtful material in ALL CAPS, LIKE THIS, BECAUSE ALL CAPS ARE SO UGLY THEY WILL REALLY JUMP OFF THE PAGE. The chances are zero that anything so obvious could survive into a final manuscript unchecked.

At the end of each day, leave a loose end to help you get started in the morning. Some people literally stop in the middle of a sentence. Others start each new day by refining yesterday's work, building up momentum as they approach the blank page.

If you don't feel like writing, write anyway. "I never write unless I am inspired," intoned Ralph Waldo Emerson. "And I see to it that I am inspired by nine o'clock each morning." This is why you journal daily, is it not? so that writing is a habit, just something you do?

Remember that you cannot tell how well you are writing by the way it feels. Most writers have an occasional day when writing feels like flying: Your mind is quick and clear, your verbs are active, the sun is out, and you feel like a minor deity, able to recreate the world (if only re- and only on paper). Then there's the occasional day that feels like creeping along a trail in the dark, extruding each sentence slowly and with pain from somewhere near the umbilicus. In between those extremes come days when you feel like a carpenter, putting together a pretty darn good bookcase.

Naturally, we all prefer to fly. But flying, in my observation, is no more likely to produce excellent work than is creeping. Either can produce brilliance, windy trash, an admixture, or good workmanlike prose, and you cannot know which till you look at it cold, later. Carpentry is less risky, though it may turn out flat.

In short, you should accept however the writing day goes, but don't get married to the results till you look at it later, cold. It's only a draft.

Remember that A DRAFT IS ONLY A DRAFT, which is why the final two chapters will deal with refining your draft and feeling stuck.

Refining Your Draft

*The first law of
intelligent tinkering
is to save all the parts.*
—Poul Anderson

Refining your draft is much like editing someone else's
work, except that you always have the writer handy—
maybe too handy, as the inner writer tends to defend the
status quo. ("Oh, but that image is so funny.")

An editor, by definition, has one enormous advantage
that the writer does not: a fresh eye. Not knowing what
the manuscript is supposed to say, the editor can tell what
it *does* say, the better to spot any gaps and goofs. Editing
your own work is hard primarily because you lack that
outsider's view.

You can approximate it, however. Don't you find that
you can often tell how something might look to someone
else? Now is the time to call on that social ability.

**Before you start refining, do whatever will freshen your
view of the manuscript. At a minimum, take a break
and print out the manuscript.** Because I revise exten-
sively, I write and print my manuscripts in galley format,
which you may care to try—single-spaced, at some forty-
two to sixty characters per line, never more. Forty-two is a
traditional line count for newspapers because that's about
how many characters the human mind can process at one
time. As a result, forty-two is highly readable. A newspaper reader runs his eye right down the middle of the col-
umn with no significant left-to-right movement, therefore no chance of losing his place. Readability is still good
at sixty characters, a width that offers the writer one extra
advantage: It puts more text on each page, so that you see
every word and sentence in its full context. Either width
gives you plenty of room to write reactions and corrections. You can scrawl whole new *paragraphs*, if you like.

Give thought as well to your typeface, because if your

text is physically hard to read, you're working under a handicap. Research demonstrated years ago that serif typefaces, the ones with high-rising l's and h's and with little cross strokes (serifs) across the top or bottom of many letters, are easier to read than sans serif types, the ones with plain letters. (This one is called Gill Sans Light.) You would expect otherwise, but so it is. Serifs make letters and words look more different from one another, so that sustained reading takes less effort. To test this notion, put a piece of paper horizontally half covering a line of the Gill Sans type. How well can you read it? Now try the same experiment with any serif type. The king of readable type, to my mind, is still Times Roman, a precomputer classic developed specifically for high readability.

While your manuscript prints, it may help to straighten your desk. Put away your writing clutter, the better to segue into editor mode. Then go do something else in another room. Even a lunch break will help, and a weekend or a week off will be better yet.

After your break, proceed as if you had never seen the manuscript before. The idea is to approximate an outsider's clear view of the piece as it stands. Take a moment to look forward to seeing the piece as a whole. (What we expect to happen tends to happen, after all.) See if you can work up an active curiosity as to how the piece will read— and make sure you will not be interrupted. The answering machine should be on and the door shut, because this reading is special. It is your first, best chance to see the piece as it is, warts, glories, and all. Permit no distractions.

Read at cruising speed, like any other reader, but jot down your reactions in the border. Note that word—your *reactions*, not fixes. Work on paper, with the computer turned off. The paper looks different from the screen, where you've seen those words so often. Reading on paper, then, will reinforce your hard-won sense of newness. Second, with your computer turned off, you'll be less tempted to go in and fix just this one little thing ... which can easily turn into three little things and the spinning of wheels. Most of us, given the chance, can spend five to ten minutes moving a single comma in and out, in and out, which feels like progress because the screen is always clean. By working on paper, you avoid such loops of vacillation.

At first, you may find yourself pulled into fixing, which you should resist. Of course, it's fine to mark any typos or grammatical problems that leap to the eye, as you'll do by reflex. Just don't stop to think about fixes. Keep moving, reserving your attention for the text and your own reactions. You want to notice every slightest flicker of boredom, impatience, confusion, put-off-ness, or pleasure. Do you have an impulse to skim? To jump ahead? To laugh? Are you working hard? Is your mind wandering? Make a quick note and keep moving.

Write barely enough that you'll know what you meant, along these lines:

Waiting for story to start. What's this about? ...
Bored ...
Woke up here, comp. lab busy at midnight a good touch
LOL [laughed out loud] ...
Skimming, impatient ...
Now I get it ...
GREAT anecdote! ...
Same idea as on page 2? feels repet ...
This man very annoying ...
Why so much detail? Don't see why matters ...
Boring ...
Huh? I thought he was dating the Marilyn Monroe! ...
Snickered ...
What happened to the baby?
This is new? Sounds like common sense ...
Feels repetitive ...
smiley-face ...
Feels jerky ...
BORING ...
Huh? Inhaler bad for asthma?
Who he? ...
Annoyed by all the first person ...
What's Jones say? Thot he the authority ...
How example relate to point? ...
Great quote ...
Interesting, but not clear ...
What happened to the baby? Getting v. impatient!

And so on. An occasional wavy squiggle down the border can serve as an all-purpose indicator that something awkward needs reworking.

**Read the text out loud, or at least murmur it to yourself,
lips moving, in order to spotlight any awkward patches.**
Are there places where you want to draw a breath but the
sentence will not allow it? Give it a wavy squiggle. Do the b's
and p's and awkward syllables begin to pile up in a way that
is uncomfortable to speak? Ditto. Does your tongue stumble,
for any reason or no reason? It's a problem. Reading out
loud brings any such problem to the forefront.

**Noting positive reactions is a must, and not only to pre-
serve morale. Most of us tend to think of editing as "fixing" what is off. We forget the other half of the job, and
maybe the more important half—retaining and strength-
ening what is good.** The better to retain it, mark it.

**When you have read straight through and are ready to
edit, continue to work on paper.** It is quicker, easier, and
more effective than working on computer because you will
often be adjusting passages in relation to each other. Scrolling,
scrolling, scrolling, up and down, back and forth—*without ever
being able to see the separate passages side by side*—is the hard way.

Your old printout being heavily scrawled with reactions,
you will need a fresh one on which to write your fixes. Try
printing it on paper of a different color, so that you can distinguish the manuscripts at a glance.

First, three general rules:

When in doubt, throw it out. If I had to choose one single
idea as my sole teaching, it would be this one, a maxim gen-
eralized from a grandmotherly bit of wisdom about dirty
laundry. "If it's doubtful, it's dirty."

On the same principle, if you fear that a word or a sentence
or a passage may be tedious, overwritten, unclear, irrelevant,
sentimental, needlessly offensive, or whatever—it is. When
in doubt, throw it out. At the least, put it in the bone heap.

**Your subconscious is your friend. If your subconscious
made you do something, ask yourself why.** Whatever mis-
take you have made, your subconscious had a reason—
maybe a good one. See if you can figure out what the problem was, a process that often feels like having a dialogue
with yourself. You ask the question and wait. In time, an answer comes drifting up:

Q. So why *did* we drop that story in there?
A. Well, it seemed to connect.
Q. And does it?
A. Well, . . . yes! Actually, it does, or part of it does, in such-
and-so a way.
Q. We'd better spell the connection out!
A. Yes, and cut the other part.

Or you might find you had felt a need to keep the human
side of science more in sight. Great! Knowing that, you can
now rummage through your notes to find a livelier, shorter,
more relevant tidbit. Or you might have been postponing the
hard work of grappling with topic X (sigh).

And so on. Once you know clearly what problem you
were trying to solve, the solution is often obvious.

**Do not follow rules, even rules promulgated in this book.
Do something intelligent.** There are no rules for writing, or
at least, no rules that are universal.

An engineer friend once asked me how I could know when
my work was good enough. I said I didn't know, that I just
did the best I could. "Oh," she said, "I couldn't stand that. It's
so ambiguous. With a building, it either stays up or it doesn't."

Are you like my friend in preferring firm ground? Hmm.
You probably would like some rules. The truth is, however,
that writing is inherently uncertain, even science writing.
Your best bet is just to keep asking: What do I want to say?
Am I saying it? Is it working?

When it isn't working, do something intelligent.

**In editing, your initial concern should be structural. Aim
to strengthen and balance the whole. Sweep through from
beginning to end, again and again, solving the problems
that your reactions pinpoint—first the big ones, then
small ones.** Whenever your editing manuscript gets too
mucked up, enter the changes and print out a fresh one. You
may at that point want to do another reaction reading.

Let's look first at a few large questions that will need to be
thought through for each and every piece you write.

**Do you actually have an opener? Or were you merely
clearing your throat?** Initial reactions like "Bored" and
"What's this <u>about?</u>" are ominous.

Sometimes writers spend their first few pages setting context or reporting history or exercising charm, any of which can constitute throat-clearing—something, anything, that the writer has to do in order to get started. We all do it, beginners and veterans alike, but you won't want to leave throat-clearing in your draft. Look now at your first paragraph, asking: Do I really need this? Does it have substance without which the reader cannot go on? Does it grab?

Go on to the next paragraph, asking the same questions, and the next, and the next. All too often, you can drop the first few paragraphs (or even pages) with no loss.

Keep going. However long you humphed and garumphed, chances are good you will eventually come to a paragraph that makes you sit up. Oh, you think. Here it is! Yes, the scene in the computer lab! It's the essence of what the story is about. With just a little tweaking, it will be perfect.

Often both pace and tone change at this turning point, as the writer settles into a stride. Even working on your own writing, you may be able to identify the real opener by its tonal shift alone.

Does the opener still match the story as it turned out to be? Does the piece deliver on its promise? Your vision and your topic evolved as you wrote—they always do. Adjust accordingly.

Perhaps your opener promised something the finished story does not deliver, or perhaps you promised too little. As always, controlling context and reader expectation is key. If the lead promises to explain why Johnny can't read, you must come up with a sensible argument. If you promise only to visit several classrooms and see various well-regarded teachers at work, the readers will be happy with that, too—unless you led them to expect "the" answer. It can be astonishing how much a weak or limited article perks up when you scale back the promise to something the article delivers.

Even when the promise is right, a first-draft opener can feel stiff and congested, at least compared with what you wrote after you were warmed up. Can you import some of that ease into your opener?

If the opener is seriously off, don't tinker. Take a new run at it. Close your eyes, imagine the central reader, and go. If she were sitting there, what would you say? Now say it on paper.

Do you actually have a closer? Between fatigue and a desire to be done, you may have simply stopped without telling the reader good-bye.

Or you may have an excellent closer buried beneath some closing boomph, some kind of unnecessary repetitive flourish that you wrote out of sheer momentum. Throat-clearing can take place at the end of a talk, too: Sometimes we cling to the mike in case we think of one more thing to say.

If a new or better closer now occurs to you, draft it. If not, leave yourself a note and come back later. Your piece still has a ways to go.

Take a look at the passages you marked as any variant of "boring." Do you want or need the material? Sometimes writing loses its fizz because the writer is proceeding out of sheer duty: It happened or the guy said it, so we write it. But maybe it doesn't belong. Maybe you'd rather emphasize the exciting *second* half of his career, and to hell with his earlier work. The question is, If that "boring" section vanished, would it be missed? What contribution does it make? What contribution *could* it make?

A contribution need not be factual, or even intellectual. Your writing also needs humanizing detail, changes of pace, a few hearty laughs, good examples, and a hundred other things. Sometimes you and your subconscious will find that you wrote a whole section for one wonderful bit, when all you needed was the bit.

Would the passage work better if heavily pruned? Or fleshed out? Or in some other part of the article?

Is the passage boring only because it is unclear? Most things seem boring when we don't understand them. Sometimes the problem is one of scale. If the information is necessary (yet "boring"), you may need to set a fuller context, to zero in on the critical part, or to take it in smaller, more digestible chunks.

Do your examples demonstrate what you say they do? Bad examples sometimes survive from before you had total command of the subject, or because you found them charming.

How's the *shape*? As a whole, does the piece flow? Is it beginning to seem inevitable, as if the segments could never have been in any other order?

Only with all big pieces in place should you go ahead to polish your writing, a process not unlike that of a plastic surgeon treating an aging movie star: you work all over. Pat pat pat, tuck tuck tuck, here there and everywhere—it's important to keep everything in synch. If you perfect the face (metaphorically, the opener) before starting the neck and belly, the contrast will make the untreated parts look worse than they are. It will throw your judgment off.

Worse, it will prevent you from seeing systemic fixes, in which you solve editorial problems by preventing them—nipping them in the bud, often a page or more before the problem shows up. This type of preemptive repair is smooth beyond belief. You may not even touch the passage where confusion first arose; the problem will seem to evaporate.

That thought is so important I'll not only repeat it, I'll put it in boldface:

Many editorial problems are best solved by preventing them—dropping back to an earlier passage to adjust for what's to come. This approach helps all writing but is especially important in writing science. More of our readers may be struggling to follow. They need all the help we can give.

Your eventual goal is a piece of writing in which all parts support all other parts—like a tensegrity, one of those geometrical shapes of stick and string in which no stick touches another. Yet the structure is stable, held by the tension among all its interrelating parts. When your article reaches that condition, readers will find it easy to get engrossed. Every word will contribute, and no momentary doubt or question will intrude. Readers will be drawn irresistibly forward.

Let's look now at some of the less obvious reactions from page 113 to see how they might help you fine-tune your work. In the process, a few general "rules" will emerge. When they work, use them. When they do not, do something intelligent.

Huh? I thought he was dating <u>the</u> Marilyn Monroe!

One common cause of confusion is that the reader has approached the passage with a misconception,

**any misconception, that was
somehow fostered earlier in the
piece.** It's bad when the reader gets
nonplussed and has to work it out,
as in this example of stupendous in-
eptitude. (Of course you would have
identified the Marilyn more clearly.)
Worse is when the reader gets nonplussed and quits reading.

Such problems are extremely
common, flagged by variants of
But-I-thought. Even something as
small as a badly chosen verb can
derail readers down the road. So,
whenever a reaction boils down to
But-I-thought, drop back to find
and rectify the source.

Huh? Inhaler <u>bad</u>
for asthma?

Occasionally, But-I-thought arises
from misinformation or incomplete
information, not in your article but
in the public mind. For instance,
many people do not know that
when inhalers are overused, the relief they give may hide the fact that
the patient is getting worse—much
worse—and needs immediate
medical attention. Whenever you
get a chance to correct such an
item, seize the opportunity. You may
save a life.

With more ordinary misconcep-
tions, stay general. Find a good
early place in which to say that
these new findings invalidate old
ideas. Sometimes it's enough to call
the research surprising. Only as a last
resort should you actually debunk,
because to repeat an error is to reinforce it.

<u>Now</u> I get it!

Frustrated comments like this one
mark material (call it B) that should

119

probably drop back a few paragraphs to complete some earlier train of thought (A) that got interrupted.

But occasionally B works better where it is. After all, you had some reason for putting it there—maybe a compelling reason. If so, try hinting at B back at A. (Little-did-he-know is too crude but has couth cousins. Sometimes it is enough merely to acknowledge the original puzzle—e.g., "for reasons that would remain unknown for 20 years").

Once the reader knows that A's loose end will be picked up later, he can relax. He might even be spurred on by curiosity as to how this little subpuzzle turns out. Do something intelligent, whatever will work best for the particular story. **What matters is that the reader should always feel sure that any loose ends will be tied up in good time, that he is in competent hands.**

Same idea as on page 2?

Take a look. Is it the same idea? If not, rework both passages into clarity. If yes, you have a structural problem. **Repetition is always the flag of a structural problem,** the question being why you felt any need to repeat the point. Your subconscious is your friend, and your subconscious made you do that. Why?

Several possibilities: Maybe your subconscious knew that the idea—call it C—was weak the first time round, so that the reader will have forgotten. Strengthen the original C till it's unforgettable. Then you can merely *refer* to it, with no need for a full repetition.

Or perhaps part of your second C variant belongs back with the original C, dropping the repetition.

Or perhaps the second reference needs to be stronger. You might forthrightly remind the reader of important point C, mentioned in the opener and now revealed in a new aspect.

Or perhaps the full discussion, all of it, belongs at the second location. If you remove C from its first location, does it leave a gap? Will the logic still track? Hmm. No gap, no problem. You can move it.

These strategies are partly a matter of personal style. I like to braid ideas, so will tend to let a theme echo rather than create a large lump. Others prefer to deal with one matter in its entirety before they go on. Either method can work. Do something intelligent.

Feels jerky.

A jerky feeling indicates that the reader briefly lost the train of thought then scrambled back on board, somewhat breathless. Some near-obvious fact or idea is probably missing—obvious enough that in writing you took it for granted and the reader got it on the rebound. Either spell things out or offer a hint.

Snickered.

Always a bad sign, snickering often flags overwriting or more overt sentiment than modern taste will bear. Get out your pruners.

Now let's look at a few tools you can use to strengthen ideas or to tuck them in unobtrusively.

Consider reparagraphing, always, for almost any type of problem. As Strunk and White point out in *The Elements of Style*, the first and last places are the position of emphasis at every level—in the sentence, the paragraph, and the piece as a whole. This insight is news you can use, probably the most powerful single tool you have to control context and emphasis.

Use the leading edge of a paragraph to direct or redirect the reader's attention. The readers will be grateful for your slalom flags ("Turn here!"), though few will consciously notice. They will enjoy the opportunity to read without working hard. In effect, you are coaching readers in how to read your article, so that they have their full attention free for what you have to say. For example:

The technical issue had long been ...	The nontechnical reader gratefully skips to the next paragraph. The engineer wakes up. "Ah! Here's the real stuff!"
Contrary to everything Jones had been taught ...	"Oh," the reader subconsciously registers. "This must be something new since I was in school. I'd better pay attention."
[After a long, marvelously detailed paragraph about perfumery in the court of Elizabeth I] "This scent obsession started long before ... " (Diane Ackerman, *Natural History of the Senses*)	"Oh," says the reader. "Out of all that richness, I'm to focus in on the obsession."

Put anything you want to emphasize in a paragraph's caboose, the place that gives the reader her final impression (and perhaps a millisecond longer of brain time). Last place gives you a way to spotlight particular words and ideas that are critical to later understanding or that have important resonance.

Be aware that emphasis ramps up sharply as you near the

end of a paragraph. The last sentence packs a punch, and the last few words pack a *big* punch. Never squander that position on anything humdrum like "he said."

Conversely, whatever you wish to de-emphasize should go in the middle of the sentence (and paragraph). The middle of the sentence is the place for necessary nothings such as these:

however (a slalom flag to mark the unexpected)
Dr. Jones explained
built in 1942
funded by the NIH

The middle of the paragraph is the place for their equivalent in full sentences.

The middle also makes a good home for items that only a few readers will need, for whatever reason—the phone number of a clinic, or a highly technical detail, or an explanation that only the least-sophisticated will need. Keep such items short and tuck them into parentheses or a subclause, in the middle.

In that way, the key reader can surge on by at full speed.

I gave that last sentence a paragraph of its own because it is so important. Think about it. By the way you paragraph, you can tuck in extra material to serve subsets of readers, yet keep the primary readers rolling. If your touch is light and selective, the effect will be one of layering and enrichment.

The various readers are vaguely aware of one another, and if their own path is clear, I believe they are reassured to see that you are taking care of everybody. Once again, you seem trustworthy.

Whenever you are tired is a good time to sweep through looking for easy, near-mechanical corrections, like those that complete this chapter. Fine-tuning many different passages so that they support one another takes a lot of thought. It's tiring. Taking out passive verbs is easy. Every once in a while, easy is good.

Drafts can lose as much as one-third of their length, and three-quarters of their tedium, by simple, mechanical pruning. At times, as with a manuscript that originated as a transcript or if you were extremely tired when you wrote, mechanical is the place to start.

Replace passive verbs with active ones, as you surely know how to do. The rule is universally taught.

The single major exception comes in science writing, where you will occasionally need to say, "A is associated with B." That's pretty darn passive. It could mean almost anything—and from a scientist's point of view, the ambiguity is the point: "A is associated with B" explicitly means that while A and B connect, no one is sure how. It might turn out that ...

A causes B, or ...
A causes C which causes B, or ...
Both A and B are caused by C, which remains to be discovered. Or ...
A and C together cause B, and C remains to be discovered. And so on.

In interviewing, it will pay you to dig into these associations. Occasionally, a scientist will pussyfoot at length to make it clear she means that A is associated with B, not more. Then she will go on to talk as if A definitely causes B.

"Tell me more" are the words that should tumble from your lips. Sometimes she does indeed know that A causes B, for reasons that will take you to the heart of the subject; she was pussyfooting as a reflex, or because her article is not yet published, or because she thought the material was over your head. Sometimes she's only pretty sure that A causes B, again for interesting reasons.

Take out the garbage words—at least, most of them. By "garbage words," I mean puny all-purpose modifiers such as *very, really, rather, sort of, kind of, somewhat, quite, absolutely, extremely,* and on and on. These words have a legitimate use in speech, as a quick way to add emphasis or shift a meaning. They do well enough, helped out by gestures and facial expression. And besides, in conversation we always get a second chance. If the other person did not understand us, we can try again.

In writing, however, we have to say it right the first time. "It's very pretty," for example: Did the writer mean that the item is attractive or insipid? The reader can neither tell nor ask.

As you look at your first draft, you may find dozens of garbage words on every page, for which you should be grateful. How useful! They flag all the places where you intuitively knew you had grabbed the wrong word but went on anyway. Cut the garbage and choose a more specific noun or verb.

Take out redundant qualifiers—adjectives and adverbs that duplicate part of what the noun or verb already means. Some are clichés, while others reveal a writer working too hard. You will know what to drop by asking the key question—*Is there any other kind?* Of "old tradition," for example, or "toxic poison"? Cackle, then fix. More examples:

unique individuals
is a *present* concern
intertangled complications
prune *down*
hang his head *down*
the *whole* idea, the *whole* town, the whole anything
traditions *of the past*
exactly pinpoint
a friend *of mine*
a sense of *relative* proportion
driven *primarily* by her feelings
fit the bill *perfectly*

Find every unaccompanied this, that, these, or those and insert the missing noun. Your prose will sound more literate, while you also clean up the train of thought. Now and then, "this" will turn out to mean something fuzzy like "everything I've talked about so far" and you *have* no train of thought.

This is an example . . .

Take out anything portentous. Portent is a miniature form of throat-clearing, often found as a "transition." Portent is best omitted: Go straight into what you want to say. Incidentally, I did not make up the following examples.

Few can forget man's exploration of the moon during the Apollo 11 mission when Neil Armstrong reported back to earth, "Houston, the Eagle has landed."

The problem of drug abuse is a topic often reported by the media and discussed by readers, listeners, and viewers alike.

**Go through and systematically look for generalizations
and judgments. If they are unsupported, insert the
who/what/where/why/how/when that made you think
so.** That done, you will often find you can drop the general-
ization, making the passage that much more lively. Specifics
often imply the general, though never the reverse.

**Look for abstractions, many of which will be flagged by
words ending with -ation or -ization. Where possible, rephrase the idea in terms of people, using active verbs.** Be
especially vigilant in science writing, as abstractions are epidemic in scientific journals. After you've read a dozen journal articles, they may sound normal to you, too.

Even after all these years, I sometimes go through a draft
explicitly looking for these tenacious little monsters. A few
sample fixes:

Utilization of the system has been low.
Few customers use the system.

A very different pattern of response emerges when subjects
    with prefrontal brain damage . . .
People with prefrontal brain damage respond very differently.

Patient population
People who use the Jones Hospital emergency room

Drop wordy clumps or replace them with single words:

so obviously important from a survival point of view
so obviously important to survival

feel a vague sense of uneasiness
feel vaguely uneasy

increased sharply in amplitude
spiked

significantly reduced volumes of gray matter
significantly less gray matter

An early clue came from an event that occurred about 150 years ago.
An early clue came about 150 years ago.

Wherever you find the word "not," look for a stronger way to state the idea. Not is weak and easily missed, so that careless readers may miss the point. Notice how much more the language pops when "not" phrases are flip-flopped into a definite form.

It was not long before
Soon

Jack Spratt couldn't eat any fat.
Jack Spratt could eat no fat.

He did not like to
He disliked
He avoided

Unpack all overpacked phrases and sentences. For example:

According to Paul Brooks, her biographer, Rachel Carson, a scientist with a literary flair, started the ecology movement with her book, *Silent Spring.*

Concise is good, but that's overdoing it. You can move some of that material elsewhere in the paragraph—and do feel free to create a new sentence.

According to biographer Paul Brooks, Rachel Carson started the ecology movement with her book, *Silent Spring.* Carson was a scientist with a literary flair . . .

Take out or prune every item of which you feel particularly proud. At least, view it with suspicion. The feeling you should have as you read your manuscript is that *Yes, that's exactly what I was trying to say, and it leads directly and smoothly to the next point.* Remember that writing itself is secondary, a tool with which to express the primary—meaning. Only other writers should notice the high caliber of your writing. The reader should be absorbed in the *content.*

If you feel actual pride, therefore, you can be almost sure that you are standing between the reader and the material. Maybe the topic pushed a hot button and you've unloaded—written polemic. Maybe you lost track of your reader and zoomed off on a tangent. Maybe you are showing off, or

detonating intellectual fireworks to distract yourself and the
reader from an emotional issue. Most likely, you've only
tried too hard, which is fine. (It's only a redraft.)

Overwriting does less harm than timidity, by and large. It
is good to experiment, good to go out on a limb, good to
play with words and ideas. If you go overboard out of sheer
exuberance, the worst-case scenario is that you've left yourself a big menu of possible deletions.

If you are suspicious of a passage yet see no place to
prune, it may be there's no problem. JUST LEAVE YOURSELF
A NOTE AND GO ON. People who are recovering from timid
writing may experience using a new and powerful technique
as showing off or overwriting.

If you are avoiding emotional issues, you will get published anyway. The price you pay is that you limit your
range, closing off both the highest and the lowest octaves. If
you will not go to dark places, you can never write as well as
Peter Matthiessen.

**Whenever your manuscript gets so battered and scribbled
that it's hard to work on, enter all changes and print out a
fresh copy. Then take a break and repeat the reactive reading.** You will be pleased at how much better it reads already.

Are you ready to let a friend see it? Or even an editor?

Outside reactions can be disconcerting, because by this
time you feel secure with the manuscript. To you, it reads
very well. It is dismaying, then, when an outsider reports
back and you see that he has missed your central point.

Any shock that big will be rare, but whatever you get
back, you should be prepared, just in case, with the attitudes
we've been rehearsing: It is only a draft. All that matters is
the final result. Thank heaven someone found that problem
before the piece went into print—a happy event that will
happen sooner than you think.

When You're Feeling Stuck

Nonwriters often say to me, "Oh, it must be wonderful to write so easily. I can just tell it's easy for you."

I'm sorry, but no. Stories go around about professional writers who write easily, but I've never known one and certainly never been one. For every easy-sounding paragraph in this book, several awkward-sounding versions were written, rewritten, written again from a new running start, and generally struggled with. One whole chapter got pitched out.

Often the hard part is less the writing than the thinking, but everyone meets a hurdle from time to time.

What to do depends on your particular variety of stuckness, which is why this chapter is written as a Q&A. The ideas may be more clear, however, if you read straight through.

Is the problem not your writing but you? Ask yourself if you are tired, hungry, angry, or thirsty. Are you fighting a headache? Do you need a break?

Sometimes writers fuel up on caffeine and adrenaline and bulldoze forward, all stops out, for hours and days at a time. The condition is oddly pleasant and sometimes unavoidable, given that the writing trade runs on deadlines. When the adrenaline runs out, however, stop. Learn to recognize what that feels like (for me it is a particular edgy queasiness), and stop to take care of yourself. Eat. Nap. Take a shower and change your clothes.

It is true that if you keep going, you will get a third wind (you already had your second), and even a fourth. You will achieve miracles. You will die young. (Just kidding. Well, sort of kidding.) The traditional writer's way of the all-night bash, legendary at the old *Life* and *Time*,

abuses the body's stress response, while the traditional
writer's death is to be found slumped over the keyboard.
Don't you think there must be a connection? ... which is
why I suggest that you not make a habit of the bash. I wish
I hadn't.

**Is the piece so rough that you cannot see what you have
beneath the surface jumble?** This problem is familiar to editors, less so to writers. Still, you might see it if you tried to
write too early, by talking to a tape, or when very tired—a
text so garbled that you feel impelled to pitch it. For example,
here is a chemist talking about impediments to learning
chemistry:

> The first one—I haven't read anything about it, most of it
> is intuitive—is how symbols become tyrannical and intim-
> idating. I'll move back and forth. I'd like to give you an ex-
> ample of that. I'm going to start off with showing you a
> few symbols to terrorize you. Then I'm going to show you
> what the symbols mean. [Shows slide.] Here's something
> plus something else equals something or another. These
> look pretty formidable, and they really are very simple.
> This is the story of symbols: They really are very simple
> ideas that just code for a very simple idea. Now what this
> means is, this is Phoenician, this is hieroglyphic Egyptian.
> But they both represent two apples plus two bananas equal
> four fruit. But when you see the symbols, they are so in-
> timidating. If I had written it this way: $[2a + 2b = 4c.]$
> you might not have been so frightened.

Before we pitch this mess, why not see what we can salvage,
starting with a simple cleanup?

> The first impediment to learning is that symbols become
> tyrannical and intimidating. I'm going to start by showing
> you a few symbols to terrorize you. Here's something plus
> something else equals something or another. They look
> formidable, but they really are simple. This is the story of
> symbols: they are just codes for very simple ideas. Now
> this one is Phoenician, this is hieroglyphic Egyptian. They
> both represent two apples plus two bananas equal four fruit.
> But they are so intimidating! If I had written it this way: $[2a
> + 2b = 4c.]$ you might not have been so frightened.

Yes, now I see an idea lurking, and we didn't even have to work very hard. I merely nipped out throat-clearing ("I'll move back and forth") and garbage modifiers ("*pretty formidable*"). We could go one step further, if we like, moving elements to clean up the train of thought and eliminate repetition:

> Symbols can become tyrannical and intimidating, even though they are merely codes for simple ideas. For example, here are some Phoenician symbols and some Egyptian hieroglyphics, both saying that something plus something else equals something or another. They look formidable, but really they both represent two apples plus two bananas equal four fruit. If I had written it this way—[$2a + 2b = 4c$]—you might not have been so frightened.

... but for pure evaluation, we wouldn't have to. A quick and dirty cleanup can do wonders.

At one magazine where I help out, we call this approach "looking for the pony," after the optimistic boy who received a steaming pile of manure for Christmas. When he saw it, he clapped his hands and plunged in with enthusiasm.

"Why are you so pleased?" asked his parents, surprised.

"*Well!*" said the boy, digging deeper. "Look at all this *stuff!* There's got to be a pony in here *somewhere!*"

As there frequently is.

Does the piece or some part of it have no apparent organization? Make a printout and go through it, paragraph by paragraph, asking yourself what you meant to say. What is the gist of each paragraph? Write a one- to three-word description beside each one.

You may find a few paragraphs with no gist; pitch them. And you may find a few long paragraphs with three or four topics; label each part.

Once you have each paragraph or group of paragraphs labeled, cut them apart and assign the chunks to categories in whatever way makes sense. Ponder the categories. How are they related? Should these two piles really be one? Should this one really be two? How does the material want to be? Do you have all the pieces? (You may need to skim through your notes and look for some segment you forgot.) Do things start to cohere once put in chronological order? Or is there an overarching theme (that must be explicated first), which gives rise to A, B, and C (the delta)? Or do A, B, and C arise separately and then converge (the watershed)? Do you see one locus that can serve for both entry and exit, so that you can use a spiral construction? Is there an intriguing question to raise, then evidence A, evidence B, conclusion F, etc., etc.? You get the picture—think. Look for the *shape*, as discussed in chapter 4.

put us

If the piece is long and complex, it can help to write a couple of sentences summarizing each pile before you try to put the piles in order. Then write out your head, subhead, and three to five key ideas (probably one key per pile).

Keep shuffling the pieces till you've got something that works, at which point you're ready to start splicing the units of thought together. Where a chunk is missing, scrawl a quickie version on a piece of paper and tape it in place. You think you'll remember the connection you intend, but by this time the subject is probably swirling in your head. So give yourself a break: write things down.

Cut-and-paste can be fun, and it is helpful even when the draft is fairly good. To this day, I find it helps to *see* the pieces, all together, all at one time, if necessary trailing across the floor.

Have you lost touch with your reader? Periodically, ask yourself the familiar questions: What does the key reader want or need to know? What items are important to the auxiliary readers? Then look to make sure you are addressing those issues and those people.

Straying from the path can be hard to spot at the time, and no two people lose focus in precisely the same way. I tend to get think-y and to wander off into intellectual thickets. At the other extreme, some writers get grooving on human detail and forget to etch in essential background information. Do you know what your particular struggle might tend to be?

Assessing the readers was discussed in chapter 4, beginning on page 69.

Are you trying to write a term paper? The closer you are to school, the more likely you are to be writing a term paper, purely out of habit. I remember the first article I ever wrote: It was about the campus cops of Cornell, and it went through four separate, from-scratch drafts before I stopped finding one more way to make it dull.

Fundamentally, in a term paper you *tell*. In professional
writing, you *show*. In a term paper, the reader is the teacher,
who by definition knows all and must read the paper anyway. In professional writing, the reader knows nothing and
must be enticed to read.

So: Are you writing sections of plodding background,
stuff a teacher might want to see that you know? Are you
striving to be complete and to work in all the appropriate
general statements?

You are not in that universe anymore.

You must now aim for accurate but just enough (whatever
that means for the particular readers), placing evocative, interesting, and newsy parts to the forefront. Leave out the
general statements so beloved by the teachers who taught
them to you; rather, build your writing like a sculpture, fit-
ting together chunks of solid observation, fact, and reason-
ing. Report phenomena—what you saw, heard, smelled, read
in a letter, felt in the air on your cheek, until no reader can
help but join you in the particular world of words that you
are sharing.

Have you been reading too much academic prose? Learned
journals are full of passive verbs, which can infect your writing if you've spent too long on research. To counteract the
spell, read several pages of some prose that has the tone you
seek just before you sit down to write. Even a few pages of E.
B. White, perhaps on the death of his pig, Fred, will be
enough to jolt passive constructions out of your brain.

**Have you fallen in love with a major character of your
story? Or in hate?** Either way, you will feel uneasy as you
write. Most of us know when our judgment is off, and so
will the reader. Readers lose confidence if they feel that the
writer is either ridiculing or flattering the subject.

If you fear you may be tipping in either direction, remind
yourself of professional norms. Call yourself back to being a
good reporter.

Look at the language you are using. Is it oversoft, or
overedgy? Language is so rich! Do we call that big man
burly, hefty, paunchy, potty, chesty, corpulent, square-bodied,
muscular, beefy, or gone to seed? Is his female counterpart
voluptuous, zaftig, busty, hippy, jiggly, fleshy, plump,
rounded, obese, or motherly looking? Does a long-legged

person stride, pace, trot, or walk fast? Is that big house a
barn of a house, a spacious home, a nine-bedroom manse, a
comfy Victorian, or a creaky fixer-upper? (Granted, we don't
need to refer to body weight at all, but the mild taboo spices
the examples.) Use the more neutral words in these ranges
and let the facts speak. To me, a lap pool and Japanese garden
say "manse" more clearly than "manse."

Look also at the facts and observations you are choosing to
use. The criterion should be, What do I have that will create
the richest, most accurate portrait of the person or situation
or idea? The result will probably include a few things the
person would just as soon you'd left out, along with a few
that he finds unexpected and flattering. Those reactions are
not a problem; the problem arises when you let them skew
your reporting.

So: Are you leaving out relevant material only because the
subject might find it embarrassing? At the other extreme, do
you find enjoyment, even just a tinge, in writing the unpleasant? Look at the quotes: Are you cleaning them up to a high
gloss that verges on fiction, charm gleaming from every word?
Are you cleaning them up less than usual, virtuously reminding yourself of your duty to report all the ums and uhs?
Head for the high ground of fact and normal practice.

Does the topic bore you? The best cure for boredom is to
find out more because, as discussed in chapter 1, anything is
interesting once you take the right approach.

Your best bet is to go back to the researchers and try to
elicit a story—an old-fashioned narrative, with a beginning,
a middle, and an end. Ask "Why this research in particular?
What caught your attention?" Unlike What's-important-here,
which you have undoubtedly asked already (as you should),
"What caught your attention?" may give you the beginning
of a narrative.

A balder way to say it: "Tell me the story of this work—
how did it begin?" (Perhaps they had a question. Perhaps
they observed an anomaly. Perhaps an advisor chose the
topic.) And what happened then? . . . and then? . . .

And a third possible tack: "Was there any point during this
research when you were surprised? Tell me about it."

In the mental realm, story is the universal solvent. It is
how our minds organize the world: This happened, then that
happened, then all of a sudden! . . . And we're hooked. We

have to know what White Buffalo Woman did next, or Frodo Baggins, or King Lear, or the third little pig. And we writers need to tell science the same way.

Is the piece too long? Length is a frequent problem, as most articles and essays today are far shorter than comparable work of thirty years back. Writers simply have to compress their words. It's painful.

A few questions that may help:

Have you done a full mechanical pruning (as described on pages 124 to 128)? Get all the small stuff; it may be enough.

Do several of your examples do the same job? Pick the best and drop the others. Sometimes one memorable story can do two jobs; when the second task arrives, you refer backward.

Are there whole sections not germane to the central topic? Out. If they are fascinating, give them to the editor as possible sidebars. No editor in his right mind will pass up a tasty sidebar.

Have you written a lot of background working up to the central topic? Drop it or condense it, taking only the key portions. Imagine your reader running for a train while you try to brief him on the first third of the piece—Quick, what do you say? Yes. Keep that part.

A worst-case scenario: the piece is three times as long as the space you have! Do not attempt to prune, as it cannot be done without losing the inviting texture of the piece. (Who cares if all the arguments remain, elegantly condensed into single sentences? It will be so dense that the reader will quit.) You have three choices: Look for some single piece of the text that will stand on its own (and that the editor likes), argue for more space, or find a market that wants the detail.

Do you have a story idea or only a topic? If you are wandering around in a subject, either unable to put two paragraphs together or unable to shake the feeling that you're writing a term paper, you may have a mere topic, a naked noun.

If your working head could be summarized as Everything You Never Wanted to Know about Whatever, you definitely have a mere topic. Look for its verb—something changing, something happening, some kind of action. Look for the story.

Chapter 2 discusses story ideas and how to find and develop them.

Are you trying to be original? That way lies trouble. You are standing in your own way, distracting yourself by being self-conscious. Go back to basics. Who are the readers? What do you want to say to them?

When you faithfully give your readers the best that you can muster, you will be original, as discussed in chapter 1. You can count on it.

Have you told all your friends all about it? . . . and the story went stale before you could write it? That's why I urged you not to dine out on your research, back in chapter 3. But that's okay. It's hard to believe until it has happened.

To rescue the piece, your best bet is go back now to your very early notes and look for whatever struck you at the time. With any luck, you made notes on your notes—exclamation points, question marks, asterisks, or even WOW! Though some of that early material will now strike you as obvious, you can be almost sure that a good lead and grand finale are there, because human beings like to perform. Most of your sources were "up" for the initial interview. They were speaking as vividly as they ever do, and they were addressing the obvious questions that you knew to ask and that the readers will be wondering.

Have you not talked to anybody? As discussed on pages 66 and 67, a serious professional conversation with another writer or an editor can help a lot. I sometimes don't know what I think till I've bounced my ideas off other people, and often my colleague's questions and comments open up an aspect of the story that I'd overlooked.

Also, the part of me that is sitting back and observing will notice what I do: "Interesting," says the internal voice. "You told Susan that anecdote, not the one you used in the opener. Maybe you wrote up the wrong one."

A professional conversation is different in kind from a social one: You are working, not entertaining. Your goal is to see the story freshly, not to have a nice lunch. Your colleague may have little to say but questions, and she is thinking with you, not telling her own stories or sitting back in social mode. You should come away refreshed, charged up, full of new ideas, and eager to write.

It helps to take notes, because the conversation will often ramble in ways that hinder memory.

When you feel stuck, you will seldom be ready to show the manuscript. But perhaps you can show your list of major items, several sets of head-and-subhead, or several alternative openings. As you talk them over, one of them may grow legs and start to run—aha! *There's* the story!

Whatever you get from the session, you should use right away. Short-term memory degrades by about 50 percent in twenty-four hours, so at the least you'll want to jot down your new ideas immediately. Better yet, start to work with them on the day of the talk, while you're still excited. Otherwise the new ideas will evanesce, cast off as wrong or irrelevant by your older mindset, the one that had you feeling stuck.

Are you in a power struggle with your teacher or editor?
Don't be. "The editor is always right."

This arresting phrase comes from Rob Kanigel, who was a freelance writer at the time he said it, twenty-odd years ago. Of course, I did a double take. *What?* This from a man who always wanted to spend all afternoon talking through all my edits, comma by comma? From a man who always thought I had violated his manuscript?

"Oh," he said, "I don't mean that you're always *right*. I mean that if you see a problem, then it's there. Even if I don't like your edit, I have to pay attention to the problem you were trying to solve."

The more he explained his concept, the less flattering it became. "I mean," he said, "the editor is *trying*. You really *want* to understand, and you're reading more carefully than the readers will. So even if your suggestion is really, really stupid, there's something to it."

That conversation still strikes me as comical, and also brilliant: It suggests the best way for writers and editors to work together, one that keeps things amiable and produces good work.

Writers—not just you but *all* writers—need editors because, by the time a manuscript is complete, the author's objectivity is exhausted. He knows what he meant to say, but he has no way to tell whether he actually said it. For example, critical details may be missing because over time they came to seem overobvious. The writer may have a cogent, even brilliant train of thought—but at a subconscious level, not explicit. At a more superficial level, words and sentences may need retouching, for reasons the writer is too close to see.

He knows that paragraph intimately, after all. It took two hours to write, and to him everything about it now sounds inevitable.

So fundamentally, the editor is a pair of fresh eyes. The core of the job is diagnosis, as in the reactive readings you learned to do in chapter 6.

Do you find it hard to see your own work freshly? Of course you do (not impossible, but hard). Then you ought to be grateful for that interfering editor who really does see it freshly, and who can point out where confusion and boredom invade your text.

If time allows, it works well for editors (and teachers, too) not to correct so much as to react in detail—to do a reaction read, at least on the first go-round. There will be time enough later for major surgery, if needed. Chances are good, however, that if the editor reacts in detail and the writer takes it as help, the writer can find far better fixes than the editor could devise. For while the editor may have fresh eyes, the writer has her own unique asset—a head full of facts, quotes, and stories that don't yet appear in the manuscript. Once she knows an example doesn't work, pff! She can bring out another. Can the editor do that? Not likely.

From the writer's point of view, the beauty of Rob's concept is that it makes editing impersonal, therefore easier to take. Edits are not criticisms. Writers don't need editors because they have failed—writers need editors because that's the nature of writing. The issues are not "I'm right" versus "No, I'm right." Rather, writer and editor can talk about things like, Does the reader really need so much detail in the opener? Does this example belong here, or does it fit better with the point on page 7?

Working that way is fun. Try it. Even if your editor has not reacted but fixed, try to see what problem the fix was aiming at. Do you see a better way? At times, you will. If you don't, be grateful. By and large, editors make you look good.

Ignorant, ham-handed editors do exist but are scarcer than writers may think. Even bad fixers may see the problems clearly.

Do you know enough? Really? Are you sure you have finished your research? You are ready to stop researching and interviewing (1) when you understand one layer deeper than you plan to write (that will protect you from writing something deeply stupid), and (2) when you start turning up the same thing again, and again, and again. Perhaps the new bits are useful confirmation, but more detail than you can use. They make the same point as material you already had.

Are you there yet? A few signs that you are *not*:

• As you review your material, do you find yourself skipping over some passages, thinking Oh that's just too confusing? To me, that sounds as if you don't yet understand the material. When you decide what to leave out, your thought should run more like, That's interesting, but (a) off the topic, (b) too detailed, (c) so-and-so said it better, (d) and so on. You should know enough to have a *reason* for leaving something out.

• Do you have a nagging feeling that you really should have called so-and-so or found reference this-and-that? Some people never feel ready. After all, the H.M.S. *Beagle* came into port in 1836; it took Darwin another twenty-three years to write *The Origin of Species*. But if you are a perfectionist like Darwin, you have long since learned at what point to ignore your own perfectionism. For the rest of us, that nagging feeling is the subconscious trying to help us out. Listen to it. Call so-and-so.

• Have you started writing, but find yourself writing the same few paragraphs over and over, refining them to a Pulitzer-worthy polish? Rarely, those few paragraphs are in fact all that should be written, in which case you should go back to the assigning editor (or teacher) and say so. More often, polishing one nugget means you don't know enough. You are huddling on an island of certainty in a sea of confusion and must launch into immersion.

Are you *acutely* lost, in a state of total confusion and sinking fast? Every once in a while, novice writers really do tackle something that is beyond their abilities—at least, the abilities they had when they started. Perhaps that is what has happened to you now. You are growing, and a painful experience it is.

If you were sitting in my office today, asking for help, here are the questions I would ask you. Perhaps you could find someone to have the same kind of conversation with you. First question:

What is your deadline? All too often, people do not ask for help until too late, the day before (or of) the deadline. Too bad. I would've loved to help them. Question number two:

Do you have a reprint from a scholarly journal about this work? If so, start there and go through it sentence by sentence, paraphrasing each unit of thought in your own words. If you were sitting in my office, that's what I'd make you do, the idea being to find out precisely where your understanding failed—and surprisingly often, that would be all the help you needed.

Sometimes people decide that they cannot do something: understand physics, let us say. Then, when they have to do it, they can only sit there in agony looking at the pieces of paper—agonizing but not progressing, because they have so little hope that they never actually come to grips with the material. They are deer in the headlights.

Paraphrasing for someone like me helps them learn that they *can* figure it out, because I won't let them off the hook till they hazard a guess—which is right, mostly, or almost right. If it's almost right, we look up each unfamiliar word until they *can* produce the paraphrase. Great! And on to the next. And so we go. It may be only minutes before the penny drops. "This isn't so bad! I can *do* this!"

If you are afraid of your subject, is there someone who could help you in that way? You may know more and understand better than you think. Puzzled writers are often missing one or two key concepts, ideas so big that nothing makes sense without them, but not many in number. Once you locate the gap in your knowledge, you are almost home.

Lacking someone to sit with you, you can push through alone, though it's harder because all the will power has to come from you. If there is a press release, you will have to lean harder on it—and don't forget that whoever wrote the press release should be willing to answer questions. It's her job.

Even lacking a press release, however, you can go forward, especially if your article is to be fairly short. Keep looking up every technical word that came up more than once in your interviews. Make a vocabulary list and consult it as needed.

After you've gone through all the notes, ask yourself, What seems to be the main idea? Put it in your own words, two to three sentences worth.

Ask yourself, What seem to be the main three to five ideas

or links in the train of thought? Summarize each one in your own words, just a sentence or two. Great! There's your outline. Put the ideas in some sensible order and write your article. The next one will be easier.

If you go through all that looking-up and still feel unsure about your grasp, pick the most teacherly of the people you talked to and show that person your outline. Probably you're okay. If not, the teacher-person will help you untwist the last few tangles.

An article with such a history might persuade you to make an exception about showing copy. Of course, you'll show it to the teacherly person.

Most people enjoy helping anyone who will make good use of the help. You must not take someone's time and then not write the article—you would feel like a jerk, and the would-be helper might agree. If you will follow through, however, do not hesitate to ask for help. Every single person who has ever accomplished anything has had lots of help, especially in their early years.

Are you working too hard? Many of us learned in school that writing was somehow special and difficult, requiring an outline and a great many rules. The outlines used roman numerals

I. for the main idea
II. for the secondary idea, then
III. A.B.C.s and
   a. a.b.c., and at each level you had to have ... was it three? I no longer remember the details of Mrs. Richardson's nuisancy notions—though it's obvious that programmers do, because beginning with that first I., my computer kept providing a roman numeral every time I hit Return.
   b. And now it's insisting on a.'s and b.'s, which might have been handy fifty years ago, but now I need to get out and back to regular text.
   c. H-E-L-P!
   d. Mrs. Richardson lives on, the soul of a new machine.
   e. In Mrs. Richardson's classroom, all paragraphs were to begin with a topic sentence.
   f. All sentences were to be complete, with both a subject and a verb.

g. Subjects and verbs were to agree.

h. Infinitives were not for splitting.

i. Writing was not for enjoying.

After schooling like that, it's no wonder that many of us tense up when we sit down to write. But fortunately, the How of writing is not the heart of the matter, and we all once knew it.

Since you are reading this book, you were probably one of those children who loved words and ideas and wrote for pleasure—birthday cards, skits, letters, web sites, and more. These childish productions can get elaborate. I wrote a shoebox full of miniature books, illustrated in crayon, printed in grimy pencil, and bound in colored paper. A friend, as an eleven-year-old, invented a town replete with mayor, town council, and baseball team. He then spent many happy hours writing the town paper.

Were you also a writing child? I will bet you were, if not always, then at some period. Whenever it was, it will help you now to remember what fun you had. Writing can be a form of play, and when it is, the readers always know, because they are having fun, too.

Do you think you could collaborate with your inner child? Let the child write the rough draft and the adult handle deadlines and grammar.

Do you have an emotional agenda you are not revealing?
You may remember that in chapter 2 I told you not to write about the subject closest to your heart, "meaning material that came to you as a revelation, a bolt of lightning that lit up the entire internal landscape." I argued that you were too close to the material to manipulate it and would likely write with the tone of some unfortunate person ranting on the subway—as I did myself, when I tried to write about my heartfelt topic. Wait, I said. Let it season in the basement of your mind. With so many wonderful things to write about, why zero in on the one that will be the very hardest? Or, if you must write about it, why not wait till your skills are fully developed?

If you could not bear to wait, however, so be it: Your best hope is to come up front with your agenda and make an ally of the reader, as Andrew Solomon did in The Noonday Demon, an Atlas of Depression. Earning the National Book Award is not so shabby, right?

Blazing emotional intensity cannot be hidden from readers, because they are smart. They will smell the lie and distrust what you say. Your only hope is not to hide it.

Is it possible you've not actually been working? Here I am remembering the classic story that Bob Armbruster, an editor friend, used to tell about a freelance writer. This writer had a story long past deadline but was hopelessly blocked. He could not write and could not write and could not write until finally one morning his wife presented him with a mug and a thermos of coffee. Lunch was on a tray, the answering machine was on, and she would take the children out for the day. Now nothing could interfere with his writing. And she left, slamming the door.

That evening when the family came home, all the silver was spread on the dining room table and the writer was hard at work, polishing forks. "It began to bother me," he explained.

Writers always laugh at this story with a certain explosive quality, I think because it so perfectly describes something we've all done.

So: Have you been silver-polishing?

Any activity counts as silver-polishing that is just worthy enough to let you stop writing with a straight face. Cleaning your office, organizing files, returning phone calls, or defragging your hard disk are especially good because you can delude yourself that they are a necessary precondition to writing and that therefore, in fact, you *are* writing, sort of . . .

Are you trying to make the work perfect? If so, that might explain why you are silver-polishing. As long as something is not yet written, the possibility remains that it may be perfect, or so one is apt to feel.

The classic advice for perfectionists is, "<u>Don't get it right, get it written.</u>" In other words, force it. Just write: Bang something out.

I have used this method and it feels awful—until the next day, when I would arrive at the office and see that, actually, the work was not so bad. So I'd patch and polish (always fun, the exercise of craft), then start a new segment on that momentum. The next day I could start by polishing yesterday's rough edges, and so it went. But to achieve that rhythm, you must first get something written. You must begin.

At the other extreme, you can limit your writing time, which is my present approach, since I no longer go to an office. I tell myself I am going to write for one hour and then stop. I set an alarm, and when it rings, I STOP—but only in theory, because I get engrossed. I don't want to stop, which is the intended effect. It works like those marriage counselors who instruct a couple in various ways to stimulate each other, which they are to practice but under no circumstances have sex. Of course, they come back the next week with sheepish grins ... It seems to be a fact of human nature: Not having to do the deed makes it possible to start.

On occasions when the alarm rings and I'm not engrossed, I do quit, a welcome relief. Yet the single hour is enough to keep the project cooking in the back of my mind, so that in a way I am always writing, even when I might appear to be out in the yard pulling weeds. While one-hour stints do not work for research, they do for writing.

On the conceptual level, try this idea: Writing is like baseball in that what matters is the batting average, not the individual at-bat. Face it: Not everything you write will be great. In fact, some will be terrible. So what? Forgettable stuff gets forgotten. What people will remember is your good work.

If you are a baseball fan, you know that home-run sluggers have low batting averages. They mostly slice that mighty bat through empty air, while the ball goes whistling by— much like writers who aim at perfection and publish rarely. If your temperament leaves you any choice, I suggest you see yourself as an ordinary hitter, one who just tries to get on base. Sometimes you succeed, and sometimes you don't. Either way, you keep swinging. And if you keep swinging, every now and then you will hit one out.

Along the same lines, try a musical image: Did you know that Bach wrote 255 cantatas? Two hundred fifty-five! Not to mention all his masses, sonatas, concertos, preludes, and fugues. Bach was a working stiff, churning out music for the church as fast as a composition every week. Some of it does not survive and most pieces are never performed. Today, we hear only his works of genius, of which we have so many because he wrote a little something every week.

When you do not like what you have written, don't worry about it. Twenty years from now, no one will remember, not even you. Keep on writing.

Afterword

This book contains very little career advice, but I do have three final thoughts that I hope may be nuggets for novice writers.

You will probably write in much the way you handle other parts of life. If you do projects in a great bash at the last minute, surfing on adrenaline, that's the way you'll write. If you are methodical, always getting everything done well before it's due, that's the way you'll write. And so on.

It pays to adopt a certain realism in these matters, because the writing life comes in many flavors, and you need to pick the right one for you. Specifically, bashers need something to trigger their adrenaline. So they often thrive as staff, goaded on by the rest of the group (not to mention the boss). Gregarious types may not need goading but wilt in a life that leaves them alone with a keyboard all day most days.

People who do well as freelancers, by contrast, tend to enjoy their own company and also to be extremely well organized. It takes both discipline and foresight to live on an income that comes in fits and starts. Freelancers also have to keep appropriate records of every expense, like any other person running a small business, which some people find so nitpicky that they cannot remember to do it.

Are you a prickly and independent sort, easily crossed? You will want to stay out of public relations, where one is expected to write with tact. But for a more accommodating nature, writing science for a university or hospital can be very satisfying. It pays well, too.

A word of caution: Writers of any sort can easily move from journalism into public relations, but it is almost impossible to go the other way. The habit of being accommodating and tactful can leave a lasting stamp on a person's prose. Those who hire know it.

If you do not own a copy of *The Elements of Style* by William Strunk Jr. and E. B. White (Macmillan, 1959), buy one and read it with respect. It is elementary, yes, but only in the sense that atoms are. Strunk's admonitions are building blocks that you ignore at your peril.

I used to read *The Elements* straight through about once a year, to inspire myself and refresh my writerly rigor.

And finally:

Do not be so "realistic" about marketing yourself that you distort your unique development. Bringing your talent to maturity is a bit like gardening: Yes, one must water and fertilize and weed. There is work to be done. But the work pays off biggest when the plants are right for the soil and microclimate of the particular garden.

Is there some subject or writing style that comes naturally to you? Pursue it. Garden your own garden, not someone else's. And welcome to the tribe of those who struggle to write with joy and precision about our astonishing world.

Index

A graduate of Goucher College (B.A.) and Harvard University (M.A.), Elise Hancock was for many years editor of the prize-winning *Johns Hopkins Magazine*. A bimonthly, the magazine is much like *Smithsonian*, except that its articles draw from events and research at Johns Hopkins rather than from the "nation's attic."

Hancock retired in 1996, went back to school for a new master's degree, and is now a licensed acupuncturist in Baltimore, Maryland. "I can't imagine how acupuncture works," she says, "but it does. I hope to live long enough to hear a good Western explanation, which may not take long. Biomedicine is getting subtler every day."